The Mobile Poultry Slaughterhouse

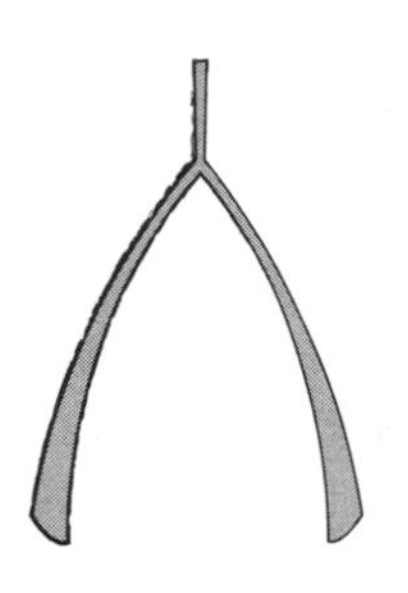

I got chickens in my front yard,
what they do is scratch and peck.
Come supper time I'll go out there
gonna find me one and wring
its neck.
It ain't meanness ya'll it's just
hungry's what I am.
There ain't nothin' in this world better
than fried chicken yes ma'am.

— Chorus to "Chickens," as sung by Hayes Carll
Written by Ray Wylie Hubbard and Hayes Carll

The Mobile Poultry Slaughterhouse

Building a Humane Chicken-Processing Unit to Strengthen Your Local Food System

Ali Berlow

Foreword by Temple Grandin

Storey Publishing

The mission of Storey Publishing is to serve our customers by publishing practical information that encourages personal independence in harmony with the environment.

Edited by Deborah Burns
Art direction and book design by Cynthia N. McFarland
Cover design by Lauren Dreier

Page illustrations by © Scotty Reifsnyder
Ephemera on pages 8, 29, and 63 courtesy of the Island Grown Initiative

Indexed by Samantha Miller
Excerpt,page ii, "Chickens," by Ray Wylie Hubbard and Hayes Carll, used by permission; and page 94, from "What Did I Love?" by Ellen Bass, first published in *The New Yorker*; from the collection *Like a Beggar*, to be published by Copper Canyon Press in 2014, used by permission

Storey Publishing
210 MASS MoCA Way
North Adams, MA 01247
www.storey.com

Storey Publishing is committed to making environmentally responsible manufacturing decisions. This book was printed in the United States on paper made from sustainably harvested fiber.

Printed in the United States by McNaughton & Gunn Inc.

10 9 8 7 6 5 4 3 2 1

LIBRARY OF CONGRESS CATALOGING-IN-PUBLICATION DATA

Berlow, Ali.
The mobile poultry slaughterhouse / by Ali Berlow.
p. cm.
Includes index.
ISBN 978-1-61212-129-1 (pbk. : alk. paper)
ISBN 978-1-60342-884-2 (ebook)
1. Slaughtering and slaughter-houses. 2. Poultry plants.
3. Poultry—Processing. I. Title.
TS1960.B47 2013
664'.93—dc23

2012043070

This book is dedicated
to all creatures great and small,
and especially feathered.

CAST OF CHARACTERS

HERE ARE THE KEY PEOPLE who played pivotal roles in establishing our mobile poultry-processing trailer. I'll refer to them often in this book. They are listed alphabetically by last name.

RICHARD ANDRE: Farmer/owner, Cleveland Farm, West Tisbury, Mass.; Poultry/Meat Coordinator for Island Grown Initiative (IGI)

STEVE BERNIER: Owner, Cronig's Market, Vineyard Haven, Mass., where Permit #417 locally grown and humanely slaughtered chickens were first sold to the general public

ROBERT BOOZ: Chef, hunter, food writer, and farm-food activist

GORDON HAMERSLEY: Chef/owner, Hamersley's Bistro in Boston's South End; cookbook author

MARINA LENT: Chilmark, Mass., Board of Health agent

JIM McLAUGHLIN: Owner, Cornerstone Farm, Norwich, New York; consultant, processing equipment distributor

JEFFERSON MUNROE: Poultry farmer, The GOOD Farm, Tisbury, Mass.; head of IGI's first Chicken Crew

• • • • • • • • • • • • • • • • • •

FREQUENTLY USED ACRONYMS

BOH: Board of Health

CAFO: Confined animal feeding operation

IGI: Island Grown Initiative

MPPT, MPPU: Mobile poultry-processing trailer or unit

USDA: United States Department of Agriculture

CONTENTS

FOREWORD

by Temple Grandin

I GET ASKED ALL THE TIME: How can you care about animals when you are a designer of slaughterhouses? People forget that in nature, every living thing eventually dies. That is the cycle of life. Many people today are totally separated from the natural world, and don't know it is harsh as well as beautiful. They visit national parks to see the beauty and do not experience the harsh reality of a predator killing its prey. Death in a well-run slaughter plant is much kinder than death in the wild. In the HBO movie about my life there is this line: "Nature is cruel, but we don't have to be."

The animals that we raise for food deserve to have a life worth living. This is the definition of animal welfare by Farm Animal Welfare Council (FAWC) in England, and I completely agree with it. Our cattle, pigs, sheep, and chickens would never have been born if we had not raised them. It is our responsibility to give them a good life and a quick, low-stress death.

I had the opportunity to travel with Ali Berlow to a small beef plant that was doing everything right. I worked with them to make a few modifications on their chute to provide a nonslip floor, which may be a minor issue for people, but for cattle, it is essential. Cattle panic when they slip. Details that people fail to notice are important for animals. The cattle remained peaceful and calm, and that is the way slaughter should be done.

Today many people in urban areas want to get back in touch with where their food comes from, and this desire to reconnect with what we eat drives many local and sustainable agriculture programs. I recently visited a Mother Earth News Fair where people could learn how to raise animals, grow crops and vegetable gardens, and process meat. I am amazed at how many people attend this type of conference.

Ali has been working to develop sustainable local poultry programs where neighbors can process their own chickens. Her book on mobile poultry slaughter facilities provides step-by-step instructions on how to start and run a local small-scale chicken-processing enterprise. Her emphasis on getting a community of producers to work together is key to success.

PREFACE

I AM A HOUSEWIFE. I am not a farmer. I have never killed a chicken, much less gutted one, nor do I intend ever to do so. However, I cook meat and I eat it, so therefore I am complicit in this cycle of life, death, and dinner.

Slaughter happens. This practical guide is about convictions and about how to make a delicate situation better. At least, that's my perspective in this book, *The Mobile Poultry Slaughterhouse.*

Cookbook for the Journey

What you have in your hands is like a cookbook: a guide and a how-to based on one community's food system and its experiences as they relate to poultry. Learning about food and what you can do about it in the face of industrial-strength barriers is part of the journey. This book is not a blueprint, but it does provide practical, helpful information about how to launch a poultry program — the keystone of which is a humane mobile slaughter option, and the goal of which is to foster a latent local food, the fabulous farmyard chicken.

Cookbooks are filled with short stories and happy endings. Recipes are interpretations. Dream big, act small, every day.

Slaughter is such a disturbing topic of conversation that many people will actually cover their ears when you bring it up. If you ever want to clear a cocktail party or silence the stranger in the airplane seat next to you, tell them you build slaughterhouses. That'll quiet them.

The more clear and grounded you are about *why* humane and accessible slaughter is so important to *you,* the more effective an advocate you will be. Be gentle with the uninitiated, though. They just want to eat in peace like the rest of us.

In my view it's time to take back the kitchen, starting with good, clean, fair food from the source. Food you can look in the eye and not betray the incalculable, extraordinary relationships between man and animal.

Look up from looking away and look around instead. Be smart. Start small, one step at a time. One chicken at a time. Use this book as you would a cookbook with recipes, stories, and happy endings, and build a mobile slaughterhouse for your own kitchen, for your own community. Build it with foundations of humanity and the welfare of the animals that nourish us. Build it from the strengths of respect and dignity, with persistence. Build with it conviction. Build it because we've got to start somewhere. The home kitchen is a powerful, creative, inviting, and inventive place. Build it because complacency is no place to cook from.

Stand and face the stove. Then turn around and get going. Cook like you mean it. Make change happen.

THANK YOU

Temple Grandin, for inspiration and a guiding light.

Farmers: What you do fascinates and sustains me.

Scott Soares, for not hanging up on me on that first phone call.

Jim and Peg McLaughlin, bless your adventurous, generous souls.

Richard Andre, Jefferson Munroe, you are the two best partners-in-crime this girl from Wisconsin could've wished for. May your knives be ever razor-sharp.

Thank you, Michael Pollan, for writing the book that started it all for me, and the Vineyard community for working together to get it done.

Holly Gleason, Walter Robb, Binky Urban, Maria Moreira, Fae Kontje-Gibbs, and Kathie Olsen for believing, making dreams come true, and having my back.

Robert Booz and Gordon Hamersley for your soulful cooking.

Lauren Dreier for creating artfully, Viki Merrick for the craft of writing, and Steve Earle for this truth: "Art is war."

To my husband, Sam Berlow, and our children, Max and Elijah, who are humane in all things slaughter and roast chicken now: You are the loves of my life.

Hey look, Mom and Paul, I wrote a book!
Love, your forever daughter and saucy wench,
Alice

Introduction

“I’ll meet you at the boat,” I said to Jim.

The day had finally arrived. Jim McLaughlin of Cornerstone Farm Ventures in New York was coming to Martha’s Vineyard to deliver Island Grown Initiative’s mobile poultry processing trailer (MPPT). It had taken almost two years to reach this point.

“What color is your truck?” I asked.

“Look for the red PT Cruiser,” he said, in a deep voice that came from somewhere near the center of the earth. Thrown off balance, I replied, “You’re kidding, right?”

“No, really,” he said. “A red PT Cruiser. Don’t you know that ‘PT’ stands for ‘poultry trailer’?”

And damned if it didn’t. On July 11, 2007, Jim and his wife, Peg, rolled off the ferry into the heady July night air of Vineyard Haven — with the MPPT in tow. “The dock workers all thought it makes cotton candy,” he said, and we smiled in icebreaking laughter that got me breathing again.

How and Why We Began

The goal of Island Grown Initiative (IGI), a nonprofit organization, is to help create and support a more sustainable agricultural system on Martha’s Vineyard. Local food includes not only crops — tomatoes, greens, asparagus — but also meat from local animals. Yet the slaughter options available to island farmers had formidable challenges. IGI envisioned a solution that would allow the animals to meet their destiny in the most humane way possible and on the land on which they lived.

From its inception in 2005, IGI asked Vineyard farmers a lot of questions, listened, sorted through the answers, raised money, and built community support for an on-island, size-appropriate, humane slaughter solution that would include both the backyard growers and the full-on farmers who want to raise local poultry.

Seeking Creative Solutions and Strategies

Vineyard farmers are certainly not alone in facing such obstacles; it is just that on an actual island the challenges are crystal clear. But in fact, there is an abyss within the agricultural infrastructure in our country: the virtual absence of accessible slaughterhouses for poultry. In rural communities across America, small family farmers who are still raising animals for commercial sale face onerous and complex barriers. Safe, clean, affordable, fair-wage, humane, USDA- or state-permitted poultry slaughter and processing facilities are few and far between. Local, state, and federal regulations and permit fees do not reflect the reality of diversified farms and remain "one-size-fits-all" — and that size is super-big. They are not scaled appropriately, nor do they reflect the multi-species specialty meats and diversified farming strategies that are used by many small family farmers.

According to the U.S. Department of Agriculture (USDA), the number of slaughterhouses nationwide declined from about 1,200 in 1992 to about 800 in 2008, and four companies controlled 58.5 percent of the market by 2007. Meanwhile, the number of small farmers increased by 108,000 between 2005 and 2010. Following the trend of conventional agriculture, the slaughterhouse business has become consolidated and industrial-size-strength, leaving the very small-scale farmer out of luck. Accessible, small-scale, humane, USDA-inspected brick and mortar slaughter and processing plants have all but gone extinct.

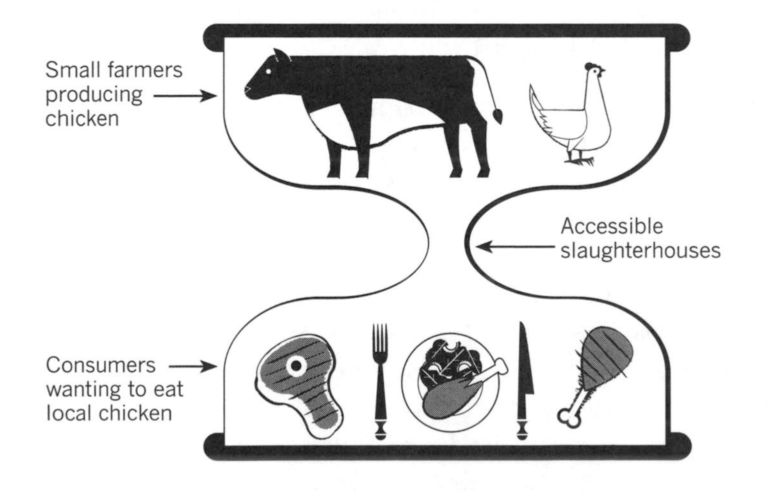

As journalist Michael Pollan and others have observed, the agricultural economy has the shape of an hourglass, and this is especially true on the local level. On top are the thousands of small farmers producing meat and poultry, or wanting to. On the bottom are the multitudes of consumers eager to eat local meat and poultry. In the middle is the bottleneck: the very few accessible slaughterhouses due to corporate consolidation.

This lack of good and accessible processing capability stifles the small family farmer who wants to raise livestock. Building a slaughter facility is prohibitively expensive without enough animals to justify and sustain it economically. But with the right support for small-is-beautiful, size-appropriate technology — a few dedicated people, some money, and a leap of faith — this can change.

Because there is so much uncharted territory in the world of mobile slaughter and processing, nothing is impossible if it's done well. It's not that anyone is strongly opposed to safe, humane, on-farm slaughter and processing that creates jobs and a good clean product for sale. It's that no one knows quite yet how to say, "Just do it."

The High-Wire Act: Juggling Multiple Stakeholders

When Jim McLaughlin landed on Martha's Vineyard in that red PT Cruiser with the MPPT in tow, island farmers were raising about 200 birds. Five years later some 20,000 broilers have been raised, humanely slaughtered, and sold, resulting in more dollars circulating locally. Today, local farmers are selling their chicken. Six part-time jobs are firmly in place and slaughter/processing fees that would've been spent off-island stayed instead in our community. In the 2012 poultry season alone, approximately 9,000 broilers were raised. This is an estimated $180,000 gross revenue (averaging a $20 price per bird) that went to farmers for the sale of poultry, minus $40,000 that was paid out to the Crew.

Managing the mini-mobile-slaughterhouse is a high-wire balancing act of spinning china plates atop tall skinny poles. It involves the animals, the farmers, the Chicken Crew, the eaters, and all the local, state, and federal agencies who in the end really don't want any of those plates to come crashing down. With strategic perseverance, those of us who wanted to bring a humane slaughtering operation to Martha's Vineyard worked with both the community and government bureaucracies to turn poultry into an economically viable option for local farmers. And it worked. One humanely slaughtered bird at a time.

A Good How-To Includes Righteous Whys

Eating meat used to be an exceptional event — for a celebration or a feast — rather than an expected, routine, everyday occurrence. These days, for the most part, people don't want to remember or deduce that their chicken sandwich or bite-sized nugget came from a living, breathing, feathered bird. And once you stop wondering, wanting to know or be reminded about the live animal, it's much easier for the cruelties of factory-farmed poultry to continue.

"Once you know, you can't *not* know," my dearest friend Holly likes to remind me. And that can result in an overwhelming sense of powerlessness, a feeling that the problems are too vast and that one person is too inconsequential to do anything about it. You don't want to know because if you did, you'd have to face the truth. One viewing of the movie *Food, Inc.* or *Fast Food Nation* (see Resources) will turn most people off to what they've already turned away from.

Even if you don't eat meat, slaughter is a part of life. And now factory farms are part of life too. Industrial agriculture, however, is a path to collapse. There are more sustainable, humane ways to raise animals and to feed people ethically, without wreaking such havoc on the environment.

Personal Tipping Point

"Ick," my neighbor said, pointing at the whole chicken. "I don't want to touch it." The glaring lights of the supermarket reflected off the plastic-wrapped bird.

"Ick?" I asked. "What's 'ick'?" My neighbor explained that she doesn't want to be reminded that chicken was a chicken — that meat comes from an animal "with a head, eyes, and feathers and stuff." Whole chickens make you recall that. She shuddered and reached instead for a package of boneless, skinless processed parts. "I like the breasts," she said. "Besides, white meat hardly has any fat." And flavor, I thought to myself, but didn't dare say it. We clearly had different takes on the topic. But there was no need to arm-wrestle food politics in front of the meat counter. Not that day, anyway.

We'd bumped into each other for a reason. This was a wake-up call for me. Thinking about our encounter, I was reminded how removed as a culture and distant from our food we've become, and willingly choose to remain. And as a food writer/activist, I saw how far down my rabbit hole I live, while naively assuming that the rest of everyone is in there with me. I realized how acceptable, convenient, and even desirable it is — this most unnatural delusion that exists between animal, meat, cooking, and knowing.

It's not that my neighbor was stupid or even that uninformed. The food choices people make are complex and not for me to judge. Instead it was a real-time study in the systems that were put firmly into place under the Earl Butz school of agriculture: "Get big or get out."

Poultry farms were some of the first to grow into agribusinesses of today. Marketing and advertising trotted right alongside, encouraging us housewives not to worry over the fuss and muss of meat. Generations of cooks could look away as processed food made cooking real food a bother. It has now been nearly 60 years of forgetting — and 60 years of factory farming that raises livestock as vertically integrated protein widgets. This mechanized, assembly-line equation of cost, time, energy, and profit cares little for animals, people, or the environment.

And my friend who said "Ick," who didn't want to touch, much less cook, a whole chicken, skin, bones and all: like the rest of us, she's just doing the best she can.

As I stood there, lulled by supermarket Muzak, fluorescent lighting, and aisle upon linoleum aisle of processed foods, I reached my own tipping point. I contemplated the array of flaccid, depressing, anonymous, boneless-skinless chicken parts — breasts, thighs, drumsticks, nuggets, and wings, wrapped in plastic and sitting on diapers and Styrofoam trays in the meat counter and frozen food section of the grocery store. How many individual birds are represented in one package, I wondered: two, four, six, or more? Was each leg from a different sentient being? And those "giblets included" — had they been repackaged separately, each organ individually, and then shoved back into the cavity of some other anonymous bird, as if they grew there in a plastic sack?

I suddenly saw supermarket chickens as a gross manifestation of disconnect, disharmony, and disillusionment, and I realized I didn't want to cook with them or feed them to anyone. I didn't want to eat these dismembered family packs, priced artificially low because of federal subsidies and political clout, and raised as protein units with no consideration for the short and sad lives of the birds. Even the full-color photographs in glossy food magazines turned my appetite off once I peered behind the curtain of how those disembodied parts were raised, how they got there, what they'd been fed, and how far they had traveled.

Once You Know

I'd had enough, and so had my appetite. "Once you know, you can't *not* know." Then the corollary arises: "And once you know, you can't not *do* something about it." Amen. Crossing these great divides of disconnects means gaining knowledge to overcome helplessness, anger, frustration, and disgust, and turning them into positive action.

So what's one thing to be done about factory farming? Even if you and I stopped eating meat, the problem would not disappear. For me, the antidote to the negatives is to do something of worth. Connect the dots in a new and healthier way. Draw functionally beautiful constellations among land, farmer, animal, humane slaughter and processing, market, and family dinner table.

Much stems from the kitchen, after all — a powerful and nurturing place. It's where we feel comfortable and where we go to be comforted. It's where I go to sort things out, and it's by far the best room of the house. Everyone seems to end up there even when there's a dining room ready to receive. But all cooks worth their salt know that their food is only going to be as good as the ingredients they start with.

A farmer can raise the hell out of a flock of chickens. She can give them a good life: fresh air, with room to spread their wings and be hen-social, to eat, scratch, and peck at grubs and bugs and grass and weeds and grain. To drink fresh water. She can protect them from raccoons, coyotes, rats, skunks, dogs, hawks, and the weather — heat waves, cold snaps, downpours, and spring squalls. Yes, a farmer can raise many healthy, happy chickens expressing their chicken-ness. But if the slaughter goes awry then it all means nothing.

There is nothing cool about killing chickens. It is a messy business. Like anything human, it can be honorable, dignified, and in accord with nature, or it can be brutal, inhumane, a butchery. An irresponsible slaughter is a horrific thing, whether it takes place in a backyard, on a farm, from a mobile unit, or within the walls of a brick-and-mortar slaughterhouse. "Local food" is not inherently better if it's not taken care of from its beginning through to its end.

Any good and humane slaughter system, whether mobile or in a building, is only as good as the people involved. I would never justify any slaughter and processing system that is inhumane or hidden from view, that lacks full transparency, or that results in unwarranted harm to animal, person, land, or food.

Postcard from an Island

Martha's Vineyard is a 100-plus-square-mile island off Cape Cod in Massachusetts. Shaped like a squashed triangle, it is a patchwork of six towns and the lands of the Wampanoag Tribe of Gay Head. It is a beautiful island of beaches, public conservation lands, rolling hills, stone walls, and open working fields with a few wind turbines, and it is a wonderful place to raise a family. There is not one McDonald's, strip mall, or big-box store.

The Vineyard's Circumstances

The media depict the Vineyard in airbrushed postcards of lighthouses and gingerbread cottages, populated by boldfaced names. The island is not all rich and famous, however, nor is it all about hobnobbing with the political and cultural elite. According to the Island Plan, Dukes County (comprising Martha's Vineyard and the neighboring island of Gosnold) is one of the poorest counties per capita in Massachusetts.

From September to May it is a rural, diverse community. An estimated 16,000 people live here year-round, working for a living, sending their kids to school, going to the grocery store the way people do everywhere. Its demographic is an ever-evolving mix based on New England Yankee, Portuguese, African-American, and Wampanoag roots. Recently it has experienced a strong influx of Brazilian immigrants.

In the summer months, the island population swells to a tightly wound 100,000, creating serious pressure on roads, movie theaters, and restaurants — and testing the nerves of the year-rounders who nevertheless depend on these tourists' dollars for their livelihood. For it is the seasonal spending during these "one hundred days of summer" that many people count on to carry them through the rest of the year.

Local Food in Demand

Despite its tony reputation, the Vineyard's culture is intricately woven with stalwart and independent fishing and farming traditions. Islands seem to attract the self-reliant and self-sufficient, and this one is no different: it is home to a thriving grassroots population and innovative local food initiatives. Since one lives and dies by the boat, the threat of being cut off from the mainland makes one think long and hard about where food comes from.

The increasing demand for local food is combined with a strong agricultural history and well-established land conservation efforts that include agricultural restrictions. As a result, a good share of the family farms and open working spaces have resisted cashing out to development pressures that could cover fertile soils with more second or third vacation homes.

Currently there are about 30 small or micro farms on the Vineyard, and about 935 acres of private and conservation lands are in food production (as reported by the Agricultural Self-Sufficiency Report, Martha's Vineyard Commission, 2010). Established farmers as well as the USDA-classified "socially disadvantaged farmers" and "beginning farmers" make up the agricultural scene, raising and selling raw milk, cheese, eggs, produce, mushrooms, flowers, meat (from beef to rabbit), and poultry. The number of backyard growers producing more of their own food is also on the rise, in line with national trends.

Given the Vineyard's high profile and high summer population, the island presents unique marketing opportunities for its small, agriculturally diverse community. It's comparable to urban farming, in that direct sale and wholesale markets are all within a 15-mile radius. The natural boundaries of shoreline and saltwater encourage interesting solutions in an evolving and resilient local food system. Strategies and solutions such as the MPPT can be case studies or models for rural communities anywhere. (See Mapping Your Own Island, page 38.)

Getting On and Off the Island

There is no bridge or tunnel to Martha's Vineyard; traveling off-island is generally by car ferry. The actual crossing of Vineyard Sound takes 45 minutes, but with buying a ticket, waiting in line, and disembarking,

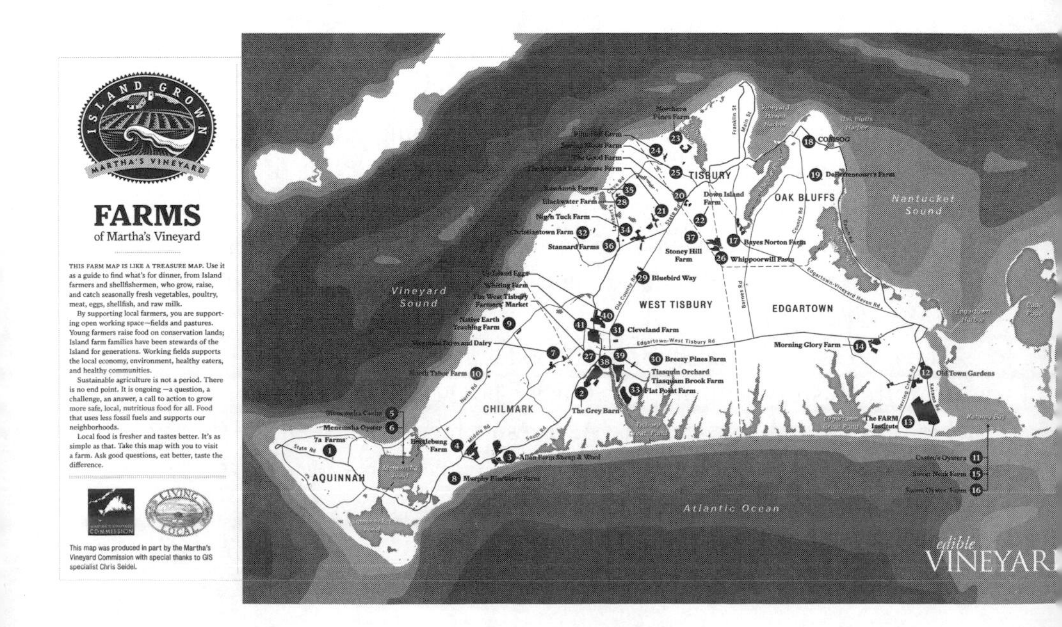

the five-mile trip takes twice that long — that is, when the winds, tides, and visibility are all favorable. The ferry shuts down or runs on a delayed, trip-by-trip basis when conditions dictate, stranding farmers, trucks, food, supplies, and livestock on both island and mainland.

For islanders, the prohibitive costs of livestock transport — boat fees, fuel, and time away from the farm — create logistical challenges that cut into already slim profit margins. It's no wonder that few farmers manage to raise animals beyond what they put in their own freezers. Some have even adjusted their livestock and slaughter cycles based on seasonally fluctuating ferry fees.

In addition, livestock (with the exception of poultry) must be crated or secured in trucks the night before early-morning departures. You don't want to miss that boat, after you scored a slaughter date at one of the densely scheduled slaughterhouses within reasonable driving distance. With four-legged livestock, you then drive as far as 250 miles to a USDA-inspected slaughterhouse. For cows, sheep, lambs, pigs, and goats, two round trips off-island are necessary: first to drop off the animals and a second time with the requisite refrigeration to pick up the frozen roasts, loins, chops, and burgers, wrapped in butcher paper or plastic. If you want bacon, a third or fourth round trip to the smokehouse may be necessary. After all that, you may be disappointed by the sorry or inconsistent state of the butchery and the quality of the cuts.

Long-distance travel is stressful for animals. Stress in turn diminishes the quality of the meat. When an animal is stressed, chemicals flood its body, causing metabolic changes in the muscles. Fear, loud noises, excessive temperatures, unpredictable motion, bad handling (catching and loading, for example) combine to reduce meat quality, color, and texture.

Market Snapshot

The Vineyard has three seasonal farmers' markets. The summer markets are popular with tourists, and the winter markets with locals, and the farmers get those direct sales dollars. There are also five year-round, locally owned, independent grocers and two Stop and Shops.

Sadly, food insecurity exists here as it does across the country. The Island Food Pantry is open three times a week from September to April. In recent years it has seen record numbers of people coming through the doors for basic provisions like canned soup, beans, cereal, tuna fish, and peanut butter.

Price Point

Local specialty or not, meat that's poorly packaged, labeled, and displayed will be one tough sell, especially when the price is higher and people are looking for good value. It may be possible to sell locally grown meat or chicken once or twice, but if it doesn't taste good at the dinner table you won't have many repeat customers.

An island-grown chicken costs between $5.00 and $6.00 a pound. It's expensive but absolutely different from a cheaper commodity chicken in many ways, including the qualities that are by far the most important to the eater — taste and texture.

We're still not meeting all the challenges in this local poultry system to make chicken reach the general market with a price point competitive to a commodity bird's. That's virtually impossible in the current system of agriculture subsidies, particularly with grain. It's nearly a miracle that we have local chicken available at all, much less in the local grocery store. (See Local Grocer + Local Chicken = True Cost of Food, page 80.) But some people are working on making this healthier meat more accessible in terms of price, and they're getting it done. (See Comfort Food, page 125). True change in local meat systems will take place only when federal policies, legislation, regulations, and direct payments shift toward better support for the small local agriculture enterprises.

A Cook's Wishbones

When two people hold both ends of a bird's clavicle and pull, legend has it, a wish will come true for the one left holding more bone. I never could bring myself to snap a wishbone that way: they're just too elegant and intriguing. Call me a different kind of superstitious, but I keep those chicken bones close. There are Mason jars of dried ones in my office for inspiration and in my kitchen for memories of meals shared. Some are turkey — big and hefty. Some are asymmetrical, lopsided even, and some are simply delicate and whimsical as if they'd blow away in a breeze. It's an amazing variety — a testament to chicken breeds and farming methods, I suppose.

Island farmers have given me a few wishbones, too. My family knows to set them aside when we eat local birds. But there's a special one that hangs on a nail in my kitchen window.

The wishbone belonging to Permit #417, the first MPPTed bird that my family ate, hangs like a relic, an homage, and a talisman to food, cooking, and the kitchen. That bone grew inside the body of a Cornish Rock Cross that had been pasture-raised and humanely slaughtered on a farm five miles from my home. That bird's waste fertilized the soil. It was humanely handled in a slaughterhouse with no walls by people who were trained and skilled in matters of animal health and welfare, and in the transforming processes that turn animal into raw and ready-to-cook meat.

I paid a fair and real price for that bird and handed my money directly to the farmer.

I paid a fair and real price for that bird and handed my money directly to the farmer. All the transactions along the way were done in the full light and full acknowledgment of the regulatory agencies in my state of Massachusetts. Nothing was done under the radar in some underground local meat black market. And there was nothing to feel but pride and dignity in this construct that fostered the land, the farmer, and the eater.

I cooked that bird from pasture to stockpot. It was one tasty, tasty roast chicken. The animal's final encore — its simmered-dry and picked-over bones — went into the compost and back to the land.

Except, that is, for its wishbone.

Permit #417 Chicken

The following recipe is for a chicken that roasts up golden, flecked with salt, ground pepper, and dried thyme. It's cooked in an iron skillet on top of a box spring of celery, onions, carrots, and lots of garlic. A squeeze of lemon, some drizzled olive oil — it's unfussy. I don't even truss the legs. Permit #417 is served with fingerling potatoes roasted with the chicken or mashed potatoes and butter, greens, crusty bread for dipping in the drippings, and a jammy red wine.

Depending on how many people I'm feeding, I can get three or four meals out of one 4- to 5-pound bird: first the roast; then the pickings off the carcass for a pasta dish, soup, or chicken salad. Finally, I put the bones in my stockpot along with an onion, a couple of celery stalks, a peeled carrot roughly chopped, a bay leaf, and a smattering of peppercorns. I'll let this simmer until I'm satisfied that the water is imbued with all the chicken-ness it can hold.

While it's still hot I pour it through a colander into a large bowl. After it's strained, the tangled remains of bone, cartilage, skin, and spent vegetables are tossed and turned into compost. I offer my dog a few tidbits. Finally, I pour or ladle the broth into Mason jars and set them to cool in the fridge. The chicken fat will rise to the top and congeal, sealing it like a lid until I need the stock for another meal.

Recipe for Permit #417 Roast Chicken

Though my family has saved a whole Mason jar's worth of wishbones in the years since the MPPT arrived, none of the chickens tasted as sweet as Permit #417. But there was no recipe for this dish when this culinary adventure began in my own backyard.

1. Start by building a slaughterhouse. You need only a small one: one that fits on the back of a 10' x 15' landscaper's trailer like a hillbilly Rubik's cube so it can be towed from farm to farm by a small pickup truck.
2. Gather your ingredients. The main one is community — your farmers, regulators, grocers, chefs, eaters. Take the initiative, form an alliance. Communicate regularly with your community.
3. Research poultry slaughter, humane slaughter, animal welfare, and local state and regulatory agencies that hold a vested interest.
4. Be transparent.
5. Stand strong.
6. Attend local food system conferences to connect with others and share ideas.
7. Acquire the equipment you need, and find and train your crew professionally.
8. Advocate for the right to raise food, to sell it, and to feed your family clean, safe, fair, humanely raised and slaughtered poultry.
9. Stand up respectfully to adversaries and nay-sayers.
10. Save your wishbones along the way.
11. Turn to page 110 on how to cook that lovely chicken. Your first permitted bird will taste the sweetest.
12. Return the bones to the soil.

1

Start Here, Get Organized

The test of a first-rate intelligence is the ability to hold two opposing ideas in mind at the same time and still retain the ability to function. One should, for example, be able to see that things are hopeless yet be determined to make them otherwise.

— F. Scott Fitzgerald, *The Crack-Up*

THE MOBILE POULTRY PROCESSING TRAILER OR MPPT, owned and managed by the nonprofit Island Grown Initiative (IGI), was developed because of these early observations:

- If local farmers in our area wanted to run a poultry slaughterhouse, or get the permit from the state themselves, they'd be doing it already and there'd be no need for any other slaughter/processing option. But they didn't, and they weren't.
- If local farmers had a commercial poultry slaughterhouse that was accessible, affordable, and convenient, they'd be raising broilers as a substantial part of their growing plans. But they didn't, and they weren't.
- Many other mobile poultry processing units (MPPUs) or mobile slaughter units (MSUs) in the country were overbuilt, meaning they were too expensive or were being underutilized by the farmers they were meant to serve because of management and training issues. Many are now shelved, or they have cost so much in outlay that it's doubtful they'll ever be financially sustainable.

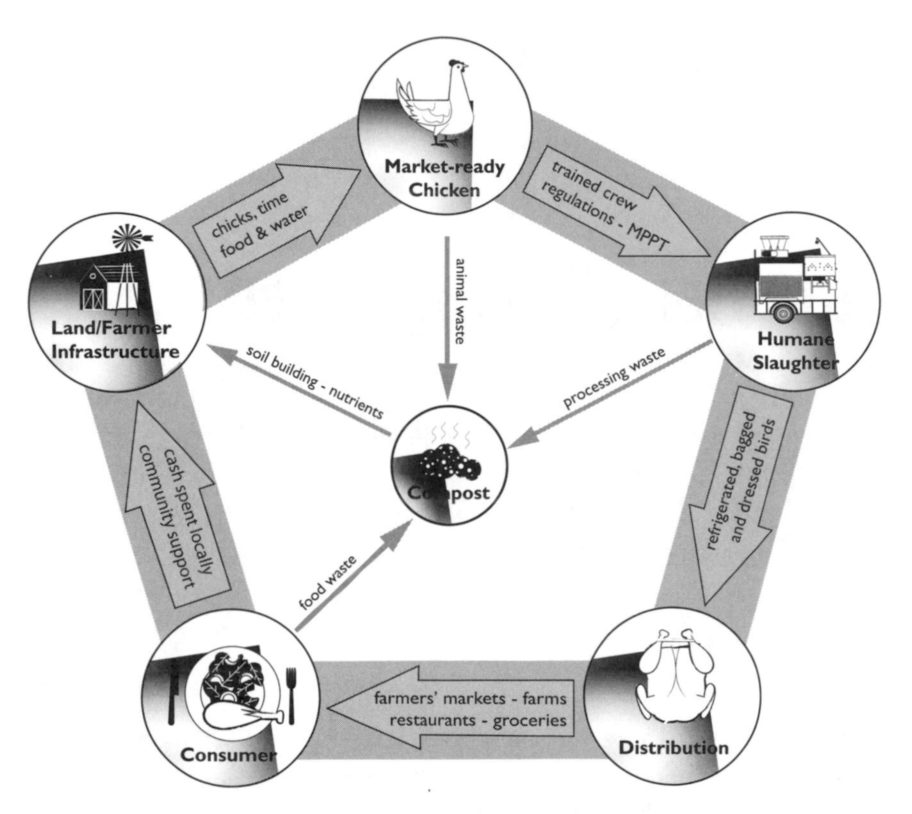

Chickens make connections in a typical food web.

Beginning Where You Are

Look at your food web from egg to roast chicken. Notice how it involves the environment, economics, policy, regulations, politics, people, birds, slaughter/processing, the kitchen, and the compost pile.

Put your local chicken at the top of a diagram. Now ask yourself, who and what make up the web around your *Gallus gallus domesticus*?

What will it take to get your chicken from the brooder to the stockpot? Consider the following. In order to support the MPPT you need:

- Land
- Farmers
- Chicken Crew
- Regulator
- Markets
- Eaters

Each of these groups has its own concerns and needs. Sketch this out to help identify what you know already and what you need to learn.

Find Your Founders

Gather people in your community who fall into the categories mentioned above to begin the discussions, dialogues, and actions early and transparently. Identify:

Farmers, who may or may not be raising meat birds now.

Markets and grocers, who hear increasing customer demand for local food, including locally raised and humanely slaughtered poultry. It's unlikely that the supply can meet this demand.

Restaurants, which have chefs and cooks looking to source good chicken. Frequently restaurateurs don't have access to local product or think it's too expensive, cumbersome, or inconsistent for their menus.

Backyard growers, who are raising chickens or want to for their own consumption. This would fall under "custom slaughter" as opposed to "commercial, licensed or permitted slaughter." These backyard growers may well become the next beginning farmers.

Eaters, who are neither farmers nor growers but want to make a difference in creating a healthier and more resilient food system.

BRING IN LOCAL REGULATORS — BUT NOT RIGHT AWAY

Regulators are part of your food web; however, get your game on a bit before you bring them into your plan. They can't be a part of your founding team, because that would be a conflict for them. And you. It's like keeping church and state separate.

These regulators work in government departments related to public health, agriculture, and environmental protection. Their concerns are:

- Livestock health
- Best farming practices
- Animal welfare
- Water
- Composting
- Food safety (including bioterrorism and traceability)
- Worker safety
- Slaughter and processing licenses

Potluck with a Purpose

Invite people from your food web to meet and discuss your community's needs. Invite the farmers, grocers, restaurateurs, cooks, parents, teachers, gardeners, writers, students, food-pantry volunteers you've identified in your food community — anyone who eats and who wants to work for a better, more resilient local food system.

Let them know in advance that this is a working meeting aimed at assigning tasks and scheduling subsequent meetings. Include possible resources for them to review before they come to your first meeting, such as Michael Pollan's book *The Omnivore's Dilemma* or *Collapse: How Societies Choose to Fail or Succeed* by Jared M. Diamond.

Serve food or make it a potluck. Always include some good, fresh food! This seems obvious, but sometimes we forget that this journey is really all about food. What you serve sends a message, sets the tone.

Have your agenda visible, whether it's on a poster on a wall or easel or on each table. Leave room for discussion — people may be shy at first — but be sure also to stick to your timeframe.

SAMPLE AGENDA

Welcome and thank you.

A reading. Start off with something from Wendell Berry, the farmer, poet, and essayist, or from someone else who inspires you.

Introductions around the room (even if you think everyone knows each other) with one or two lines about why food matters to each person. This gets vocal cords loose. It also helps you to know why people are there and what's the baseline. Listen.

Overview. What's happening in food systems elsewhere.

Discussion of the current options for poultry slaughter in your community. Monitor that this doesn't turn into a bitch session about the processors or regulators.

Determine next steps. Create actions and openly discuss how follow-up will happen: who to coordinate, how best to communicate.

Closing comments. Sum up what you heard. Announce or agree to the next meeting date, review any actions that were set, and say how and when notes will be sent out.

Clean up the kitchen, and take the slop buckets to the pigs!

Set up a sign-in sheet to collect best contact information: e-mails, phone numbers, and mailing addresses.

Use name tags.

Have fresh or dried fruit, nuts, and water on the tables so people don't have to get up for their brain-snacks.

Screen a film (see Resources for suggestions) with a discussion afterward about addressing humane raising of poultry and humane slaughter and processing of poultry.

Assign a scribe to take notes throughout the meeting.

Set up a white board or tape sheets of paper to a wall to gather people's thoughts, ideas, and, most importantly, actions. Make these notes available to everyone who attends the potluck/meeting. Share them on the Internet (such as on a listserve or Google docs) or print out copies and supply them to attendees.

Put pens and paper on each table so people can take their own notes.

Ask your group for names of other people they know who would be interested in helping but who couldn't make your first gathering. You will attract people who are passionate about food, are willing to work in hard in committed way, and want to make a difference and create change in a broken food system.

Keep the meeting lively, focused, and on point. End it when you said you were going to end it.

Follow up with notes. After the meeting, send out the notes and include the actions with names attached to them. Examples:

Find out what state agency is responsible for permitting poultry slaughter and processing in the state of ________. Patricia will do this by ______ date.

Contact poultry processing equipment distributors for pricing. Marge will do this by ________.

Create a farmer survey of questions to ask farmers in your region. Because Melinda knows a lot of the farmers, she will do this by _____.
(See Appendix, page 126, for survey suggestions.)

A nice touch to keep people's interest between your bigger meetings: send out relevant information (such as news stories) to your burgeoning group to help inform, unify, and rally your commonalities and commitment to the project.

Collaborate: Who's Out There Already?

Look around your community. Is there a local food organization, sustainability group, or Transition Group that you can connect with?

Examples include farmers' market and farm-to-school groups, sustainable agriculture organizations, community-supported agriculture (CSA) members, county planning commissions, and economic development teams.

Ways of Becoming

A mobile slaughterhouse can be a private enterprise, or, as a different model, some states own mobile slaughterhouses and run them in collaboration with nonprofits. (See below and page 28.)

IGI took on the MPPT as one of its programs. The initial investment was high, but over a few years and as the program solidified its place in the community, the intensive day-to-day behind-the-scenes work diminished.

The Nonprofit Route

Start your own nonprofit organization or, better yet, find a fiscal sponsor. Investigate nonprofits in your community that may be a good fit for your budding program and can shelter your program as a fiscal sponsorship.

Or you could file for your own not-for-profit or 501(c)(3) status. As mentioned, the MPPT described here was run by a nonprofit. As odd as that first seemed to the agencies, I believe the MPPT's success is largely attributable to the fact that it *was* run by a local, community-based nonprofit organization and not a private enterprise. IGI's only skin in the game is supporting farmers; it isn't a farmer itself, a farm, a processor, or some other related agriculture business. Trust developed.

Tread carefully, though: starting a new nonprofit is a big undertaking. With good stewardship it can be functional and fiscally responsible. Act with prudence at the beginning of your development so that you don't overreach your capacity as an organization or create expectations that aren't attainable. Avoid making promises, especially to farmers, that you can't keep. As always, be transparent along the way.

STAYING AUTHENTIC

Recruit the right board members for your organization. Strong and resilient nonprofits have a diversity of voices and depth of expertise on their board. But that doesn't mean too many people. Seven to eleven tops.

Major donors should remain and be respected as donors. They should not be officers of the board. Otherwise you run the risk that a major donor drives and controls the board and the agenda with the power of money — like the way Big Ag does in the world of political lobbying and policy influence. Diversify your board, open it up to eaters, and commit your program to transparency.

Employ from the get-go a code of conduct that includes language about conflicts of interest and term limits. Robert's Rules of Order (www.robertsrules.org), though seemingly unyielding and out of character for a small, developing grassroots organization, will help provide the structure and formalities that encourage civil discourse among the group. That way, from the most introverted to extroverted, all have a chance to speak safely and be respected by the group.

The MPPT as a Private Enterprise

So you want to run a mobile poultry slaughterhouse as a private business? Do the numbers, research the permits. Buy the equipment (see Resources), lay the groundwork with your regulators, and get the word out that you're open and ready for business. There's more than one way to skin a cat, and there's more than one way to build and run a slaughterhouse.

TIME INVESTMENT

Depending on your circumstances it will most likely take a good six months to a year to get a poultry program or the MPPT in full swing. Permitting may take longer. In the meantime, keep raising birds if you fall under 20,000 per year as a grower/processor (see Exemptions, page 87) and keep raising awareness (see Education, page 66), money (see Finding Funding, page 28), and confidence in your community.

Gathering the Players

You won't have a mobile slaughterhouse if you don't have farmers who raise livestock. What are the steps in pulling together the active players?

Find the Farmers

For a robust and festive beginning, host a Farmers' Dinner. This event is specifically for the farmers and backyard growers in your community so you can ask them their opinions, share your developing plan, and get your survey back. Invite all farmers whether they raise chickens or not. Listen hard and take good notes.

As said, transparency is key. Work openly and respectfully with the farmers and the regulators in your community. Most likely they are your neighbors. You'll run into each other on the soccer field, at the bank or farmers' market, or in the grocery store. The end goal for everyone is safe, humane food. Understand that from the farmer to the grocer to the eater to the regulator, they all approach the goal of safe chicken through different emphases and missions.

Since you come from a solid place of conviction — the right to create or support a local food system and to eat the food you trust and believe in — transparency and respect will strengthen and embolden your efforts as you proceed down the path of the permitting process (see The Path to a Permit, page 82).

Find the Chicken Crew

At the beginning of your program, you will create three to six part-time jobs because of the MPPT. Here are some steps in your hiring process.

Look for a Crew manager right away, someone with good communication and organizational skills. He or she will schedule the slaughter and processing dates with the farmers and work with local and state regulators as necessary. The manager maintains an online calendar accessible to local and state inspectors so they may see where and when the MPPT is in use. IGI pays the Crew manager a monthly stipend for these duties (see pages 26–27).

Determine pay structure for the Crew. Humane slaughter and processing of livestock is a skilled job with enormous responsibilities. Pay worthy people well (see chapter 2).

Crew wages and farmers' fees are closely tied together, of course (see page 27). Consider that a farmer must determine her processing costs as accurately as possible. IGI's cost-per-bird fee system was based on equivalent wages for the Crew and the general efficiency of a processing. IGI established this over the first few seasons and tinkered with it as the Crew faced different regulations, as well as the obligatory monitoring of paperwork throughout the permitting process.

Place ads in local newspapers for fair-wage poultry processors. Look for online listings and young farmer initiatives.

Prepare to train the Chicken Crew to run the MPPT, schedule the equipment with farmers, and maintain the equipment. (See Training the Chicken Crew, page 50.)

Why Chicken

As a nonprofit, IGI was dedicated to supporting island farmers and raising awareness about the importance of locally grown food. As advocates, we responded to the resounding chorus we heard at our Farmers' Dinner: "We need a slaughterhouse."

Four-legged, we thought at first, but upon further investigation it became clear. The best strategy was to start small, start inexpensively (manage your outlay and risk), start with livestock that's more manageable, start developing relationships with the regulators, and learn how to do it all well. Start with poultry.

Some reasons:

- An increasing number of backyard growers are gravitating toward raising chickens as an entry point to farming.
- The investment for a farmer raising broilers is less in terms of cash, time, and land than it is for sheep, pigs, goats, or cattle.
- USDA federal slaughter and processing exemptions exist for farmers raising fewer than 20,000 birds a year, should a state choose to recognize them (as the Commonwealth of Massachusetts eventually did; see chapter 6, The Path to a Permit, page 82). Add up all the broilers raised on Martha's Vineyard by all the farmers — even after an increase in production from 200 birds in 2007, to 9,000 in 2012 — and there are still fewer than 20,000 birds processed in one year.
- 7,000 of those birds in 2012 represents about 70 unique processings by IGI. And every bird sells *here*, within 100 square miles.

Questions Will Find Answers

Even when you've made an excellent plan, barriers will rise. Are MPPT systems up to food safety standards? Are the workers safe and fairly paid? Who are the managers? Who will inspect? Are you setting up your systems with a paper trail of information that can be passed on in a coherent, consistent way, or is it all in your head?

The USDA is mandated by law to inspect four-legged livestock, but federal inspectors are stretched thin. A grower/producer is required to use an USDA-inspected slaughterhouse for poultry in order to sell across state lines. Do you really need to cross state lines?

To establish a paper record, keep every e-mail used in correspondence with regulators. After phone calls or face-to-face meetings with them, write up your notes and send a copy to the regulator(s) for record keeping. This keeps all participating individuals abreast of changing situations, conversations, the clarification of terms, definition of regulations and exemptions, as interpreted and agreed upon by the stakeholders, as your program begins to grow.

Goals Will Become Clear

For us, these results unfolded over four years as we negotiated with the state to obtain the proper permits. The goal was to enable farmers to sell their chickens wherever they wished: farm stands, farmers' markets, restaurants, boardinghouses, grocers, even institutions (like schools and hospitals). As advocate, IGI worked to make all outlets available and let the farmers decide how to engage those markets. Some were all set with their own farm stands. Others wanted to diversify their paths to market. All we wanted to do was to provide safe, affordable, clean, fair-wage, accessible, size-appropriate, humane, permitted slaughter and processing. That's all, and that's everything.

The agricultural community will soon outgrow the MPPT. The demand for the equipment, the wear and tear, will soon cause it to reach its maximum capacity. And that's the point. It's a bridge piece of infrastructure. We had to start somewhere, and it had to be affordable.

So how do we move forward, collectively, serving the non-negotiable mission of safe food, with ever-decreasing budgets, onerous policies, growing consumer demand, and increasing mandates for regulatory implementation? This is the ongoing dialogue and the challenge we all face as we move toward more regional, smaller food hubs.

Crossing Cultures, Earning Trust

When our newspaper ad ("Help wanted: Chicken processors needed") didn't work, Elio Silva came to our rescue.

Elio grew up on his family's coffee plantation near Cuieté Velho in Minas Gerais, Brazil, where, he says, "the only thing we bought was salt. The sugar, cattle, pork, chicken, ducks, and vegetables were raised on the farm." He moved to Martha's Vineyard about 20 years ago and became a businessman, a grocer, and a good-food advocate for health and local economies. He is very active in the social fabric of the many Brazilian immigrants who made the island their home. He goes to church. He's a family man. He has a lean and kinetic energy that is industrious, multitasking, and multilingual.

Elio's market, which caters to Brazilians, is a good barometer of the ebb and flow of who's living on the island. He says the local Brazilian population peaked at 7,000 in 2006, then dipped to around 2,000 in 2012 as Brazil's economy stabilized. These days he's bringing in more foods to satisfy the hunger of people from Ecuador, Mexico, and Uruguay. The Jamaicans, he says, "enjoy chicken feet, cows' feet, and oxtail, and the Brazilians will pay more for pastured chicken. They have a greater appreciation for all the parts, feet, gizzards, and old hens, something Americans haven't picked up so much."

"For me it's unfair and unjust that the farmers are just surviving. They should be thriving."

Elio described his impetus for helping develop the MPPT by finding the Chicken Crew: "I saw that the MPPT was an opportunity to help farmers make money, another way to get revenue. For me it's unfair and unjust that the farmers are just surviving. They should be thriving. To have more options to come to the table and to have a good business to pass on to their kids, for now and for future generations, I want to see farmers succeed. . . . The MPPT is a way to help. It's an option for farmers so they can succeed."

Elio introduced me to Flavio and Marcia Souza. At the time, Flavio owned a landscaping business, his wife cleaned houses, and their kids went to school in Oak Bluffs. Flavio had a truck and they spoke English. He and Marcia were the first Crew. They also recruited a couple of friends, so when Jim arrived with the MPPT and held the first training, the Souzas were ready.

Reaching across the culture gap was a great move. It was empowering to work cross-culturally in a community that has had its tensions with integrating and accepting the influx of Brazilian immigrants. Flavio, like Elio, held place and stature in the community, and their presence helped strengthen the program.

IGI wanted to ensure they were paid a fair wage, but the market had to support it. We created a pay structure based not on speed but on quality. The work turned out to be a nice balance for Flavio and Marcia, for a while. They'd do a slaughter once or twice a week as they balanced their other jobs.

Between the events, which early in the program were not so frequent, the MPPT sat in the side yard of my home on North William Street in Vineyard Haven, covered with a blue tarp, the kind used for boats. The neighborhood flock of semi-wild turkeys came by every day, scratching and pecking underneath the trailer. Our kids threw baseballs around it.

Before long, though, the demands of running the MPPT were competing with Flavio's business. He became more active in his church, and in the middle of the MPPT's second year he moved his family off-island to become a preacher. He had his own flock to tend. Fortunately, IGI's Richard Andre had been mentoring a young farmer, Jefferson Munroe, and together they took up the management and leadership of the Chicken Crew and led it forward.

CHALLENGES LIE AHEAD

If you decide to build a mobile poultry slaughterhouse, be aware that you are pioneering in uncharted territory. Like every "early adapter" you are very likely to encounter resistance. You can't know where it will come from: regulators, neighbors, animal rights people, or even competing programs. You will be told that this is impossible, illegal, cruel, dangerous, dirty, gross. By persevering, however — calmly, mindfully, methodically, strategically and transparently — you will ultimately enhance the quality of life in your community.

2

Money In and Money Out

> *The other birds that you can buy in the supermarket are based on overall efficiencies. And those efficiencies are based on price point. My efficiency is based on an ethical system.*
>
> — Jefferson Munroe, The GOOD Farm

IS A MOBILE UNIT or a small slaughterhouse economically sustainable in your community? If so, what business model in this agricultural environment makes sense for today and for the future, accounting for growth? Where will you sell the product, who is going to buy it, and for how much?

At IGI we have found that the mini version, a Lilliputian MPPT, is the perfect answer in terms of size-appropriate technology to jump-start production of local poultry. In the first year of operating the MPPT under Permit #417, $80–90,000 circulated locally; the second year, $120–140,000; the third year closed out at around $200,000. These approximate numbers reflect:

- Money earned by the Crew
- Slaughterhouse processing fees that would've otherwise gone elsewhere (just hypothetical because there isn't a poultry slaughterhouse in our region)
- The sales of chickens (farmer revenue)

These are dollars that didn't exist in our community, prior to the mobile slaughterhouse.

Figuring Out Finances

If sustainability is the ongoing goal, and fair food is one of the philosophical pillars of the food system you want to create, then you must support paying a fair wage and asking a fair price along the chickens' food chain. A new program will need subsidization at first but should not be overly dependent on outside sources.

Wages

What are comparable hourly wages in your community? IGI developed a pay structure for the Crew that was commensurate with competitive local jobs, such as landscaping and catering (see box on page 27).

Though farmers think in terms of cost per bird, IGI strategized an hourly rate for the Crew, not a per-bird rate. This established a high safety standard, based on animal welfare and quality of work, instead of processing the highest number of birds in the shortest amount of time.

Once processing flow was up to speed, IGI could shift to a cost per bird without subsidies. When you remove the financial incentive for speed, you decrease potential harm to people and birds. This does not, however, create disincentives for efficiency. Speed is a path toward bad practices. Efficiency implies that animal and human welfare and food safety are the primary concerns.

Farmer Fees and Responsibilities

The farmer always owns the birds. Neither IGI nor the Chicken Crew ever takes ownership of the birds, alive or dead.

The farmer pays the manager of the Chicken Crew directly. The manager then pays out to the other Crew members.

If something goes wrong with the equipment or there is a delay in processing and the Crew is on site for an extra 2 hours, who is to pay for that time? These are scenarios that should be discussed, prior to events. In IGI's case, the onus is on the Crew manager to ensure that a processing goes on despite, for example, a mechanical breakdown of a piece of equipment. On the other hand, if the farmer is responsible for the delay, he would have to pay an added "penalty" fee to the Crew for their time.

If a farmer has fulfilled the points of the Farmer's Checklist (see page 127), however, she should be pretty well prepared, ready to go and on time, barring all unforeseen circumstances.

FUNDAMENTALS OF FEE STRUCTURE

The farmer rents the MPPT directly from IGI for a fee of $135 per event. This fee covers depreciation and administration of the MPPT plus the shrink bags, propane, and clips, which IGI supplies.

In addition, IGI has an agreed-upon rate per bird with the farmer: $3.50 for an unlicensed/custom slaughter or $4.25 for a licensed/permitted farmer. Sometimes a flat rate is determined for custom events with fewer than 50 birds.

What goes into fee structure and what doesn't?

Wages for the Crew. IGI pays the Crew manager a $300 monthly stipend, for about eight months of the year, for his administrative duties. Novice Crew workers start at a lower hourly wage (around $12/hour to start) and move up from there.

Permit fees. IGI also pays the yearly permit fee to the state. All the commercial farms that register with IGI at the beginning of the year will be covered under this "umbrella" license, as they abide by the program.

Insurance. Explains Jefferson Munroe, "In IGI's case, a local insurance company provides liability insurance. Because the equipment is rented by the farmer, IGI's liability just covers transport of the equipment. The farmers have to get their own liability insurance to cover the product that is processed."

Taking Stock at the Start

Once you have an idea of how many broilers are being raised in your community — or how many birds could be raised in your community — you'll be able to determine what kind of mobile option you need to jump-start poultry production.

This is a risk-assessment type of situation. How much money are you willing to risk on poultry that may or may not have a place in your current food system now — taking into account where, if any, slaughterhouses exist in your area, while considering the farmers?

In 2007 there were only 200 meat birds being raised on Martha's Vineyard: effectively, zero. Zero is a good number to start with. You can only go up from there. Zero also meant that we had to meet our community where it was and not overbuild. Hence, we built size-appropriately for zero-plus growth. We built a mini.

Purchasing Equipment

Buy the best equipment you can afford. When all was said and done, the MPPT cost IGI around $20,000, and that included training the Crew. The equipment itself cost approximately $15,000 in 2007. IGI incurred no major outlays for improvements or repairs in four years of running the equipment.

The self-timing, stainless steel Poultryman Rotary Scalder is worth its expense because of its steady, reliable rotation and efficiency. Otherwise a vigilant member of the Chicken Crew would have to be on scalder duty. The Poultryman Plucker, also stainless steel, is an enviable workhorse that produces a gently plucked clean bird. Have these two pieces of heavy equipment modified with welded-on wheels for easier off-and-on maneuverability from the trailer.

For more on equipment, see chapter 3. For sources of processing equipment, see Resources.

Finding Funding

State and federal funds are available to support local agriculture and the building of infrastructure such as slaughter and processing solutions for communities. There are also private foundations interested in jump-starting local agriculture via infrastructure. The IGI mobile slaughterhouse was supported by private donations from the local community — an angel donor, a building company, a grocer, a local foundation, and T-shirt sales.

Government Programs

There are government funds for programs, education, and outreach for new, beginning, and "disadvantaged" farmers. Get to know your regional USDA Rural Development office, both online and in person.

Begin by visiting www.usda.gov. Search for "Grants and Loans," currently listed under "Programs and Services," then click on "Assisting Rural Communities." IGI received a grant for educational outreach from the USDA's Northeast Sustainable Agricultural Resource and Education (NESARE) program. Find information for your region at www.sare.org.

Once you've found an opportunity, schedule a face-to-face get-to-know-each-other visit with your public servants. It can do wonders in breaking down barriers and strengthening communication and trust.

Private Donations

Ten thousand dollars were given to IGI by Scott Lively, then-owner of Dakota Organic Beef, located in Howard, South Dakota. This amount was just about all the seed money we needed to purchase the basic MPPT (not including the double-sided sink). So when you're looking for funds and donations, look into the private food sector as well.

As always, know how your donors like to be thanked and acknowledged. Many philanthropists are low-profile and prefer that their anonymity be respected — but that doesn't mean they don't appreciate (or need for tax purposes) a letter of genuine gratitude.

Foundations

To help locate grant applications from foundations, many public libraries have access to the http://foundationcenter.org database. This is an extensive database and it takes some time and skill to sort through it all. Always read up on the foundations' giving histories. Some have very specific parameters, and there's no reason to waste time and energy chasing up the wrong tree. A well-written, polite letter of inquiry will clarify whether you and your program are potentially a good match.

Putting the Fun Back in Fundraising

IGI hosted a fundraiser at the Agricultural Hall with the theme "Local Meat Is Good to Eat but There's More to Life than Chicken." A local cook roasted up some pigs, chickens, and beef. Scottish Bakehouse (see page 125) gave us a good deal on sides. A live band played, we rolled in the MPPT for people to inspect, and we sold T-shirts.

The event was affordably priced. Its purpose was to raise awareness about the poultry program and some money, too. But the long-term goal was to get the community excited and geared up for the next phase, to build a brick-and-mortar USDA-inspected facility for both four-leggeds and poultry.

Besides accomplishing all that, the event was a huge success and a lot of fun. (See also Resources for funding leads, page 128.)

Trojan Units

Early in our program, IGI was offered the use of an MSU, a big mobile slaughter unit on a tractor-trailer for four-legged livestock. The donor was enthusiastic and said she would pay to get the unit from the Midwest to Martha's Vineyard.

I thought we'd hit the jackpot. We could show it off and excite potential donors and eaters. But doubts quickly crept in. It probably would've made a great photo op with the smiling donor and smiling farmers and a local leg of lamb or side of pig cooking on the grill. But it wouldn't have done anything to bring a true and reliable humane slaughter and processing option to our farming community.

No people-infrastructure was in place then and there was no regulatory infrastructure to help us achieve our goal of increasing the number of people raising livestock. We had no business plan, no regulators on board, no USDA inspector (as required for every permitted four-legged slaughter). Certainly there were no logistics in place for a USDA-inspected facility in which the primal cuts could be stored, hung, cut and packed; few ready farms or farmers; and no community buy-in. The articles cut out from local papers would have have yellowed and become litter for the henhouse long after the MSU was driven back onto the ferry and headed for its next trick. And the naysayers would have been proved right: "Mobile slaughter is no solution at all" or my favorite, "Mobile units belong in the 'box of things that shouldn't happen.' " Indeed, mobile slaughter does not fit neatly into the regulatory box.

Regulators and Big Ag would sigh in relief for similar reasons and some different ones too. As for the nonprofit, there would have been adverse consequences from a public relations standpoint and for fundraising down the line, because the unit offered no real gain.

The moral of the story: be careful what you jump into. The gift horse, as generous and well intentioned as it may be, could very well create more long-term problems and difficulties down the line, if you are not properly prepared to accept it.

The Consolidation of Slaughterhouses

Small-scale livestock production has declined in our country because the fundamental piece of infrastructure — the slaughterhouse — is not readily available due to cost, geography, and oversized, prohibitive regulations. Meanwhile, factory farms and their counterparts, assembly-line factory-like slaughterhouses, damage the environment, workers, and the food they put into our food system.

A slaughterhouse is "a place that is no place," writes French author Noëlie Vialles. In our society we have moved the slaughterhouses out of sight, out of mind — literally and figuratively — until an undercover video showing sick and beaten animals hits the Web or there is yet another meat recall. Either way, we look away. We don't want to see behind those walls where lives are transformed into steaks and chops and drumsticks. Slaughter is disturbing and messy. It makes people uncomfortable. Besides raw milk, no topic is more likely to rile normally sane folk into a fervent rectitude, impervious to negotiation — when they'll consider the subject at all.

In our society we have moved the slaughterhouses out of sight, out of mind — literally and figuratively.

But what have we lost in turning a blind eye?

Scott Soares, the former Massachusetts Commissioner of Agriculture, remembers three slaughterhouses in his hometown of Dartmouth, Mass. In the state today, there is no commercial brick-and-mortar slaughterhouse for chickens at all. Not one.

As "a place that is no place," the MPPT puts slaughter back into full view and on the farm. The size and place of the solution are scaled down appropriately in response to the size and place of the situation. An MPPT is every place and no place: a pop-up slaughterhouse, as it were. Here one day, transparent, wall-less; then gone. And we repeat, flock by flock, one chicken at a time. The next step in this process will be to negotiate responsive regulations as well.

Rise of the CAFO

The decline of slaughterhouses is in a direct inverse relationship to the rise of industrial agriculture and the CAFOs (Concentrated or Confined Animal Feeding Operations) that define factory farming. The vertical integration and consolidation of agribusiness includes slaughterhouses that are monolithic in the sheer volume of animals processed, the assembly-line efficiencies accounted solely as profit margins but discounting people, environment, and animals. That's centralized power.

The U.S. Environmental Protection Agency defines Animal Feeding Operations (AFO/CAFO) as "agricultural operations where animals are

The MPPT is microscopic in the grand scheme of things, yet roundly magnificent in its impact.

kept and raised in confined situations. AFOs congregate animals, feed, manure and urine, dead animals, and production operations on a small land area. Feed is brought to the animals rather than the animals grazing or otherwise seeking feed in pastures, fields, or on rangeland."

A CAFO is not necessarily defined by numbers of animals or size of operation but by the conditions under which livestock are raised. It's a methodology. A system. It is possible in theory for a "small family farm" to employ CAFO methods. Labels, language, and the lexicon of sustainability (including the very words *local food, sustainability, small family farms*) are slippery at best, empty at worst, until defined. The MPPT is microscopic in the grand scheme of things, yet roundly magnificent in its impact (see Small Really Is Beautiful, page 39).

The Chicken Equivalent

The broiler's version of the steer's CAFO is the factory-like warehouse in which birds have less than half a square foot of space in which to live their lives. According to the *Humane Society of the United States' Report: The Welfare of Animals in the Broiler Chicken Industry*, "Stocking density, the number of birds per unit of floor space, indicates the level at which the animals are crowded together in a grower house. For a chicken nearing market weight (5 pounds or 2.27 kg), the average industry stocking density is slightly larger than the area of a single sheet of letter-sized paper, 97.3–118.1 inches squared (628–762 cm squared) per bird."

The broilers in this production model have also been genetically bred for rapid growth and large breasts (we eaters choose white meat over dark meat) with a fast feed-to-growth conversion (six weeks to slaughter), and any natural activity could slow production growth and schedules. Many birds die due to overcrowding, disease, overheating, and a range of health issues — thanks to heavy body size relative to skeletal structure and internal organs that cannot support life as these animals know it.

Commodity chicken — chicken that is raised in a thoroughly vertically integrated economic model from egg-to-chick-to-feed-to-farmer-to-slaughter-to-marketing-to-product — is at an industrial-strength scale that very purposefully reduces the animal to a cog, a protein unit,

SCORING THE BROILERS

The poultry industry as a whole is one of the most difficult areas of livestock welfare to regulate, according to Temple Grandin, author of *Animals Make Us Human, Humane Livestock Handling*, and other books. For producers, the more densely packed the birds are, especially with laying hens, the more economical it is to raise them. Dr. Grandin gives "market-ready broilers" one of three scores:

1: Not able to walk 10 paces

2: Able to walk 10 paces crooked and lame

3: Able to walk 10 paces normally

in the assembly line, and a sadly efficient one at that. According to the National Chicken Council, which represents the broiler industry in Washington, DC:

> "In the 1930s, the hatching of broiler chicks was spread among some 11,000 independent facilities with an average capacity of 24,000 eggs. By 2001, the number of hatcheries had declined by 97 percent — to only 323 — but with an average incubator capacity of 2.7 million eggs."

Vertical integration at this scale has the potential to reduce the farmer and the workers in slaughter and processing plants to just another variable.

Witnessing and Reporting

An entire genre of books has emerged focusing on this destructive system. In 1906 Upton Sinclair published *The Jungle*, a fictionalized muckraking of Chicago's stockyards and meatpacking lines, which Henry Ford studied to hone his revolutionary method of assembly-line car manufacturing. The book inspired others far and wide to decry the indecencies occurring in these difficult, marginalized, shielded places where we are told, "Don't look. We'll take care of it so you don't have to." Recent tapings of inhumane treatment of animals, and drives to make video recordings inside slaughterhouses illegal, show how divisive, dangerous, and empowering, depending on which side you're on, witness can be.

Many books, websites, and movie documentaries have chronicled the rise of CAFOs and their destructive forces, disease- and pollution-spreading outcomes, and impacts on people, animals, the environment, and food. That story is beyond the scope of this book, but you can find information at your library, movie theater, or independent bookseller — no matter from what angle you approach the meat bird. Whether you want to learn about the true costs of cheap food, industrial agriculture, backyard farming, the environment, food justice, history, the kitchen, or federal policy and the Farm Bill, there is a book, a film, a website to delight, enrage, enlighten, and edify — to spur your action, from flint to spark to fire.

3

Nuts, Bolts, and Values

What scale is appropriate? It depends on what we are trying to do.
— E. F. Schumacher, *Small Is Beautiful*

THIS CHAPTER FOCUSES on setting up the slaughtering unit, including the why and some of the how. First, however, an introduction to our system, which is serving as a model.

The Mobile Poultry Processing Trailer (MPPT) is the sum collection of equipment needed to humanely slaughter and process poultry on-farm. (For the complete list, see What You Will Need, starting on page 42.) For readers' ease, it will also be referred to in this book as a mobile slaughterhouse.

The Chicken Crew consists of private contractors, hired by the farmer to do the work of the MPPT and trained by Island Grown Initiative (IGI; see page 1). The manager is paid a stipend by IGI to manage, maintain, and schedule the unit with farmers and interface with regulatory agents.

This MPPT was not a privately developed, owned, and operated piece of equipment, and it was not funded by any government agency. Area individuals and businesses made private donations to IGI to build, train the Crew, and implement the unit for the farming community. Once the MPPT was in practice, IGI did receive grant money for educational support from the Northeast Sustainable Agriculture Resource and Education program, part of a larger USDA initiative. (See chapter 2.)

COMMUNITY PROCESSING DAYS

To lower the cost of processing, say, 25 custom-killed birds, consider organizing a community processing day. Backyard growers come together to a neutral location (not on an active farm) to work with the Chicken Crew to have their birds humanely slaughtered and processed in sequence, flock by flock. The IGI piloted this on the Martha's Vineyard Agricultural Society's land where no active animal farming was taking place. Author Michael Pollan even dropped by to see what was going on. (See Resources.)

A couple of suggestions: Begin by talking with your local board of health. Seek out your extension agent and/or your department of agriculture for biosecurity issues and flock inspections. Take time to organize and sort out logistics. The community day would be terrific as a regularly scheduled offering.

Coming to Terms: A Vocabulary of Values

IGI's mobile poultry processing trailer is a modular, size-appropriate, accessible, affordable, clean, safe, fair-wage, permitted, humane slaughter and processing option for people who raise poultry for sale. Let's break down all of those descriptors.

Mobile. The trailer moves from place to place. Antonym: brick-and-mortar facility — a building.

Modular. Each piece of equipment stands alone, independent of the others and unattached to the trailer. The trailer is the base on which everything is affixed for transport. It is not designed nor is it meant to be used as a subsurface like a floor, for the Crew to work on.

Size-appropriate. In the spirit of E. F. Schumacher's *Small Is Beautiful* (see page 39), this is a technology that takes people into account and reflects the context while fulfilling the need. (It's true. Small really is beautiful.)

Accessible. The trailer is available to the community that needs it: the farmers and the backyard growers. This will mean different things to these different groups depending on scale, such as processing 3,000 chickens raised to sell in a season versus a family's flock of 25 for the freezer.

Affordable. The unit is affordable to build and maintain, and affordable for farmers to rent.

Safe. The site of the MPPT must be safe for the people (Crew, farmers, visitors), the animals, the environment, and the food itself.

Permitted. Without clearance by the regulatory agency in your state you are limited in how, where, and in what form (whole bird or parts) chicken can be sold. If it's not permitted, and you sell it, you're flying under the radar, which isn't good for anyone or any food system. In fact, it detracts from and damages small food systems — something Big Ag might love. Do it right even if your state doesn't yet know how to permit it, as was the case in Massachusetts. Work transparently within the system to change it. Smart, size-appropriate regulations are part of a safe and resilient food system and they are what we need to work toward in order to change our current food system.

Clean. This quality has a double meaning covering both equipment and food. Clean equipment and impeccable hygiene standards (see chapter 7) make for clean, as in safe, food. In the local-food vernacular, though, clean food is food that's been raised without the additives or antibiotics given to commodity factory-farmed chicken. Generally speaking, *clean food* is the opposite of commodity and/or processed food.

Fair wage. A fair wage is a paid wage commensurate with the work. The work of the Crew is skilled labor, not unskilled.

Humane. The program must meet the highest standards of animal handling and welfare and provide the quickest death and least suffering to every animal. Humane keeps us human.

Slaughter. *Slaughter* is killing an animal for food. Please don't diminish this act with euphemisms. Slaughtering is not a harvest or a collection. Use the word. *Butchering* is what happens to the carcass, the raw meat. Butchery should never happen to a live animal.

Processing. After the slaughter comes the processing. This is a transformative sequence of events starting after the death of the animal that turns it into raw food.

MAP YOUR OWN ISLAND

You can picture your own area as an island in which you try to supply as many food needs as possible. What area will you delineate as your community? Consider the current resources, including farmers, restaurants, grocers, farmers' markets, and food advocacy groups.

How far away or how close to home do you want to operate your MPPT? One suggestion: School systems may be your guide. Or factor in the price of gasoline. Once you spreadsheet-out the costs and distances to travel to farms, and how many birds the MPPT needs to work with, it will become apparent what geographical area you can effectively and financially service.

If there's a community that wants to use the trailer but is too far away, perhaps they would be better served by starting their own program. Give them a copy of this book!

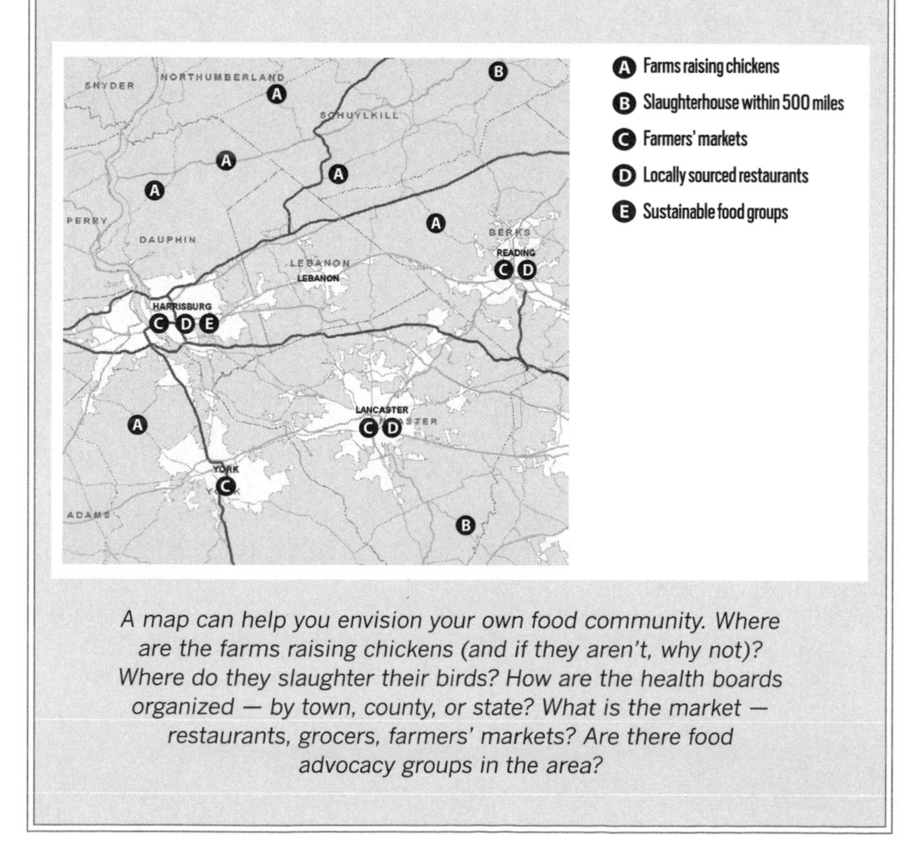

A map can help you envision your own food community. Where are the farms raising chickens (and if they aren't, why not)? Where do they slaughter their birds? How are the health boards organized — by town, county, or state? What is the market — restaurants, grocers, farmers' markets? Are there food advocacy groups in the area?

Small Really Is Beautiful

It was a summer day in 2010 when Walter Robb, the co-CEO of Whole Foods Market, made it out to see what the fuss over the MPPT was all about. We met at Richard Andre's Cleveland Farm in West Tisbury where the Crew was in the process of . . . the process.

Jefferson Munroe was on kill duty. Calmly holding each bird tucked under his arm, he extended the neck, slit its throat, placed the bird in the cone, held it through its death throes, and moved on to the next. Blood coagulated into the color of stale pudding on the stainless steel. The routine spraying of cooking oil over and in the cones prior to the slaughtering would make the cleanup easy at the end of the day.

One at a time, the birds went from hands to cone to scalder. Emerging wet they next went into the plucker. Walter and I stood by

ONE SMALL BOOK

So dog-eared, underlined, starred, checkmarked, and coffee-stained was my library copy of *Small Is Beautiful* by E. F. Schumacher that I had to send the West Tisbury Free Library a clean new book to replace it. Its spine was even broken, most notably where the chapter titled "The Proper Use of Land" opened to this passage:

> The fundamental 'principle' of modern industry, on the other hand, is that it deals with man-devised processes which work reliably only when applied to man-devised, non-living materials. The ideal of industry is the elimination of living substances. Man-made materials are preferable to nature materials, because we can make them to measure and apply perfect quality control. Man-made machines work more reliably and more predictably than do such living substance as men. The ideal of industry is to eliminate the living factor, even including the human factor, and to turn the productive process over to machines. As Alfred North Whitehead defined life as 'an offensive directed against the repetitious mechanism of the universe,' so we may define modern industry as 'an offensive against the unpredictability, unpunctuality, general waywardness and cussedness of living nature, including man.'
>
> In other words, there can be no doubt that the fundamental 'principles' of agriculture and of industry, far from being compatible with each other, are in opposition.

and out of the way, walking around and talking, as did some of Richard's free-range laying hens.

Walter is a thoughtful guy with the boyish good looks of Opie from Mayberry. He took it all in and talked with the Crew as the pile of feathers grew slowly into wet white dunes beneath the plucker. Only the intermittent rasp of the knife sharpener broke the natural calm that permeated the work of the day. The Crew and Walter took a quick break to have their picture taken out from under the tents and returned to the jobs at hand after turning up WVVY-FM on the radio.

"Appropriate technology," Walter said to me. "You've read Schumacher, haven't you? *Small Is Beautiful*?" No, I hadn't, although the title rang a bell. But once I got my hands on the library's copy, little lights went on in my head. Then I understood what Walter was talking about, how the MPPT connects land to people to animals to economics. How this pint-sized slaughterhouse on wheels stands in shining opposition to the agricultural industrialization of meat and poultry. And that is why I had to get the library a new copy.

Mobile Units across the Continent

Many mobile and humane slaughter and processing systems are springing up around North America. Here are a few examples. For a more complete list including four-legged, as well as updates on slaughter programs, discussions about the nitty gritty, and informative webinars regarding mobile slaughter units, check out the Niche Meat Processor Assistance Network (see Resources).

Kentucky

One of the first mobile slaughterhouses was built in Kentucky in 2001, four years after its inception in 1997, in a collaboration among Heifer International, Kentucky State University, and Partners for Family Farms (a nonprofit). It is also state-approved for aquaculture — caviar, paddle fish, and prawns. Since 2005 the MPU has operated smoothly and is considered a model for other communities, employing a docking station strategy. The farmers themselves are trained every two years to use it. The cost to plan and build was $70,000. SARE provided $15,000 in seed money, and the balance was raised by Partners for Family Farms and others. The MPU coordinator is an employee of Kentucky State University. All other operating expenses are covered by user fees.

Massachusetts

There are two other mobile units in Massachusetts: one open-air, one enclosed. Funded primarily by government dollars, they are projects of the New England Small Farm Institute (NESFI) and New Entry Sustainable Farming. The farmer/producer is trained to use the unit and pays the state permit fee to become a licensed processor. This is in contrast to IGI's model: training a separate Crew and holding an umbrella permit for all commercial farmers/producers. NESFI and New Entry contributed greatly to negotiating DPH requirements in the state.

Montana

"Not in my county," a state regulator reportedly said about mobile slaughter. The barriers the Montana Poultry Growers Co-op faced seem mostly due to wildly disparate interpretations of the regulations from county to county. And due to the distances the unit had to travel, the initial equipment took a beating from the road and weather. Though the farmer training manual was signed off on by regulators, the farmers

ASKING WHY

In terms of equipment requirements from your state, remember that no one really knows how to sign off on mobile units. They remain outside of the neat and tidy box of regulations. They make people nervous.

Here's one lesson we learned from looking at the few programs that exist across the country. When the regulators see your mobile unit and then tell you to go back to the drawing room to recast the equipment or the layout in some way, costing you more money, ask them politely, "Why?" How will their suggestion or new requirement, this new hoop you have to jump through, make the mobile slaughterhouse any better? It is just a pile of stainless steel and PVC piping, after all. It will be only as good, safe, or clean as the people running it make it.

Asking why makes everyone stop, listen, and discuss. Asking why may save your program thousands of dollars that would otherwise be ill-spent on refurbishments that really won't change the overall outcome of a mobile slaughterhouse's functionality or even its ability to become permitted. Only people can do that. Come together over the why to make it work for everyone.

themselves did not always know or comply with requisite training, basic cleaning of equipment, and/or keeping their local regulators informed. Hence some regulators were caught off guard.

Takeaway: start early to educate people and pave the way for the mobile slaughterhouse. They're doing it. Go Montana Poultry Growers Co-op!

Vermont

In 2008 the Vermont Legislature and the Castanea Foundation spent $93,000 for an enclosed unit to be operated by the state of Vermont so farmers could do on-farm processing of their poultry under state inspection. The original operator of the MPPU decided to pass on renewing his lease. The MPPU went to auction in January of 2012 and was sold to a local farm operation in Vermont for $61,000.

Asunder

"Watch your step. There's a spider. Don't step on it," Jim Athearn said, pointing down into the trampled grass. Spider? I didn't see a spider. "No reason to set asunder one of God's creatures."

Over our shoulders, just beyond in the back barn, Flavio and the Chicken Crew were well into the work of the flock. About 120 broilers would be bagged and ready for sale the following day. "You may think it's a bit hypocritical of me to say that," Jim went on, "as just over there we're taking lives for the soup pot."

I didn't, actually. I thought instead how thoughtful it was. Every day a farmer like Jim, an eater like myself, literally makes life-and-death decisions about food that we don't even see or know about. "God's creatures set asunder" reflect decisions that are made for us behind veils. They are actions and reactions to situations and circumstances that are mostly camouflaged, just like that spider. The difference on Morning Glory Farm was that life and death were in full view — the spider and the flock. But until they were seen beyond the veils, they would have remained obscured from sight, from being.

What You Will Need

As mentioned, IGI's little slaughterhouse-that-could will cost around $20,000. That will include all the stainless steel equipment you need to humanely slaughter and process chickens. Turkeys can be managed with this equipment too, if you're so inclined. And that's buying good equipment, new: kill cones, scalder, plucker, evisceration table and all the accompanying components. Have wheels welded to the bottom of the heaviest pieces, such as the scalder and the plucker, so you won't hurt your back trying to heft them on and off the trailer.

Below is the general equipment list for a mobile poultry processing trailer. Buy the best equipment you can. Do not skimp. It is an investment that will serve you well. Here's a list of what you will need, down to the duct tape.

BASIC SETUP

- ☐ Two 10' x 10' pop-up tents. Brand preference: E-Z Up Tents
- ☐ 6 stainless steel kill cones; Lazy Susan style is what IGI's mobile slaughterhouse uses, but watch sharp edges.
- ☐ Self-timing rotary scalder, modified with wheels
- ☐ Plucker, modified with wheels

Kill cones

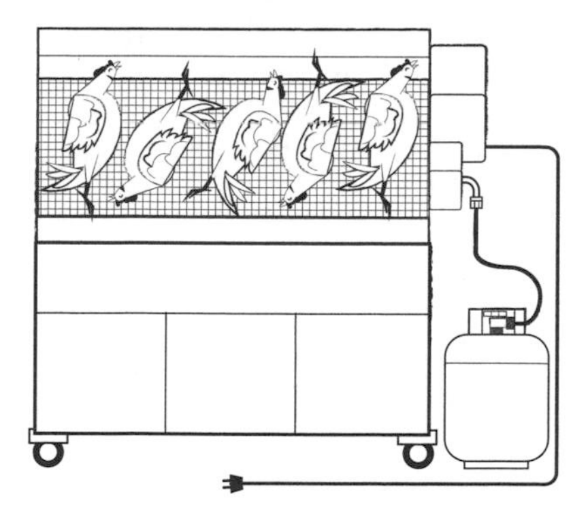

Self-timing rotary scalder

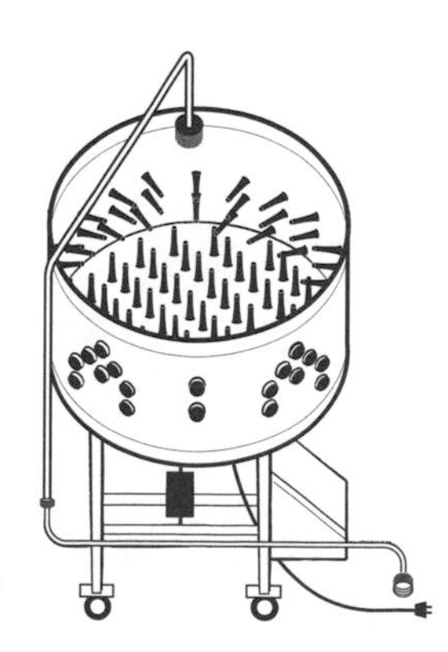

Plucker

(list continues next page)

- ☐ Stainless steel or plastic evisceration table with custom railing and funnel hole cut in the middle (see Resources for Poultryman products)
- ☐ Two 20-pound propane tanks
- ☐ Double-sided hot and cold water sink with hand soap and paper towel rack (see page 46 for building instructions)
- ☐ Six plastic 32-gallon chill tanks for birds, marked in indelible ink **EDIBLE**
- ☐ Three 3-gallon plastic chill buckets for edible offal (hearts, livers, gizzards), marked in indelible ink **EDIBLE**
- ☐ One 5-gallon plastic chill bucket for feet — marked in indelible ink **EDIBLE**
- ☐ Two tanks for compostable materials — marked in indelible ink **INEDIBLE** (one for under evisceration table, one for catching blood at kill cones)
- ☐ Plastic table for surface to place paperwork, bagged birds, etc.
- ☐ Drain rack (see page 59)
- ☐ Large pot for bagging (size for deep-frying a turkey or boiling lobster) and accompanying burner
- ☐ Knives and scrapers
- ☐ Electric knife sharpener (Chef's Choice 130 is a nice one)

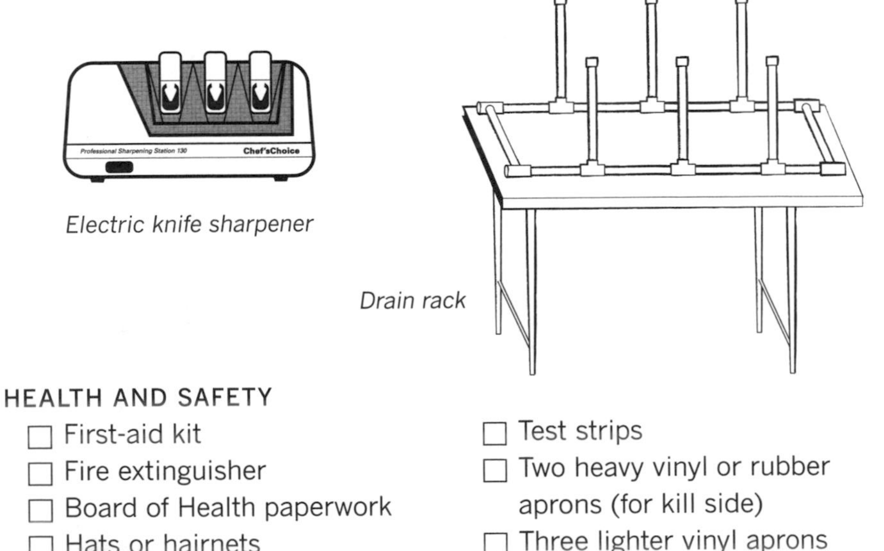

Electric knife sharpener

Drain rack

HEALTH AND SAFETY

- ☐ First-aid kit
- ☐ Fire extinguisher
- ☐ Board of Health paperwork
- ☐ Hats or hairnets
- ☐ Bleach (no additives, scents or colors)
- ☐ Test strips
- ☐ Two heavy vinyl or rubber aprons (for kill side)
- ☐ Three lighter vinyl aprons (for evisceration side)

IMPORTANT ACCESSORIES

- ☐ Shrink bags, clips, pliers, labels
- ☐ Dish soap for scalder water
- ☐ Spray-on cooking oil (such as PAM) for kill cones
- ☐ Bleach-water mix for sanitizing (see page 95)
- ☐ Clothes hangers for aprons
- ☐ Two or three spray bottles for sanitizing
- ☐ Green scrubbies and 5-gallon buckets for cleaning
- ☐ Leg grabber
- ☐ Colored zip ties for temperature monitoring
- ☐ Two thermometers, preferably at least one digital
- ☐ Portable radio for Crew
- ☐ Duct tape

POWER AND WATER LINES

- ☐ Two 100' 12-gauge extension cords
- ☐ Two 50' 12-gauge extension cords
- ☐ Two 25' 12-gauge extension cords
- ☐ One 100' potable water hose (often found at marine supply shops)
- ☐ Two 50' potable water hoses
- ☐ Three 25' potable water hoses
- ☐ Three pistol grip spray heads for hoses
- ☐ One four-way splitter for hoses

SETTING UP THE HOSES

Here are the six hoses you need and where they go:

1. One hose from spigot to splitter
2. One hose from splitter to plucker
3. One hose from splitter to evisceration table
4. A second hose for the evisceration table
5. One hose from splitter to sink
6. One hose to increase the distance from the spigot or to use filling chill tanks or for cleaning up.

Double-Sided Hot and Cold Water Sink

Save yourself about $3,000 and build this double-sided sink. A double-sided sink is an integral part of an effective barrier system between the "kill side" and the "evisceration side" of your setup. A physical barrier is essential to help protect a freshly killed bird from contamination.

This design is based on a sink IGI used for several seasons that was approved by boards of health. Some advantages of this unit include larger wheels, larger capacity for wastewater, less electricity required, and no need to refill the water heater.

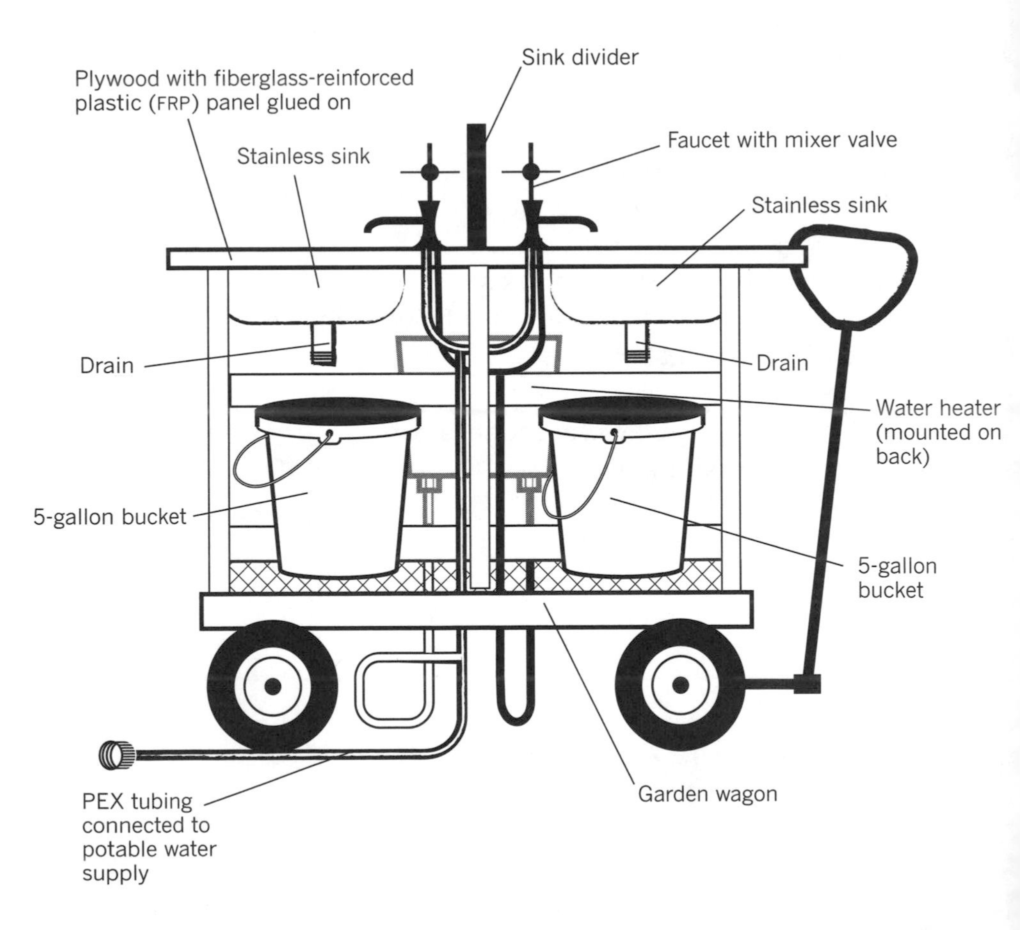

SCHEMATIC DRAWING OF DOUBLE-SIDED SINK, FRONT VIEW

devised by Jefferson Munroe

PARTS

One 4' x 8' piece of fiberglass-reinforced plastic (FRP)

One 4' x 8' piece of 5/8" plywood

One garden cart — preferably with a metal screen bottom so water will drain out

Three 8' 2x4s

Four carriage bolts with washers and nuts

Two 5-gallon buckets

Two stainless sinks with faucets and mixer valves

Two door hinges

½" PEX tubing

Three ½" PEX T fittings

Six ½" PEX ½" pipe fittings

One ½" PEX female hose fitting

One Ariston 6+ gallon six-year 1500-kW, 120-volt point-of-use mini electric water heater (available at Home Depot)

STEPS

1. Glue the FRP to the plywood.

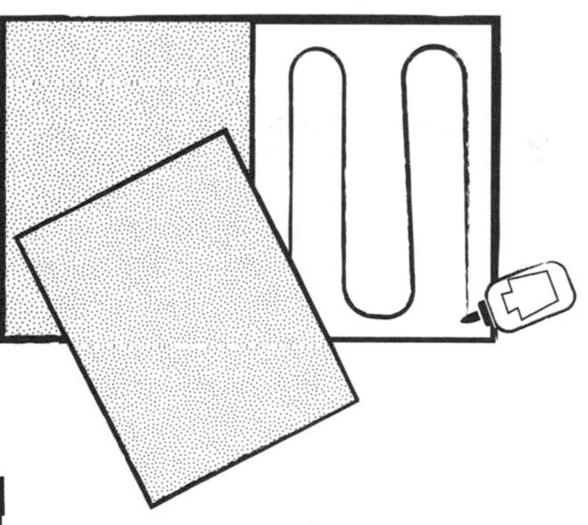

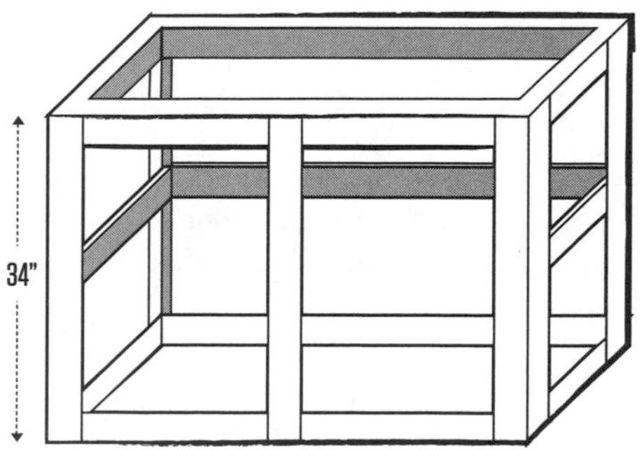

2. Put together the wagon, then build a box out of the lumber to reach a height of 34".

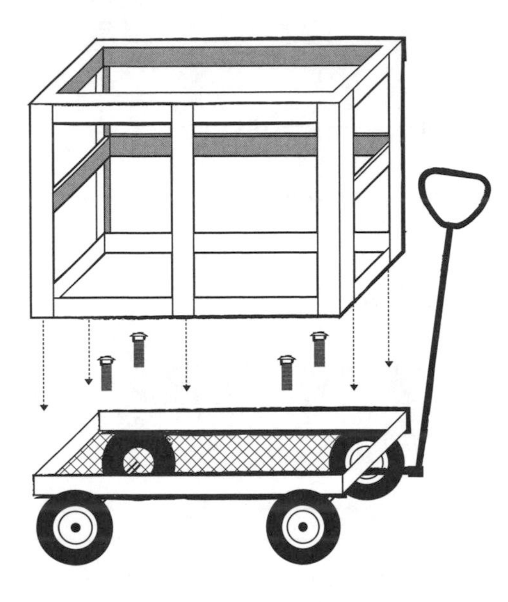

3. Attach the box to the wagon with carriage bolts through the metal grate.

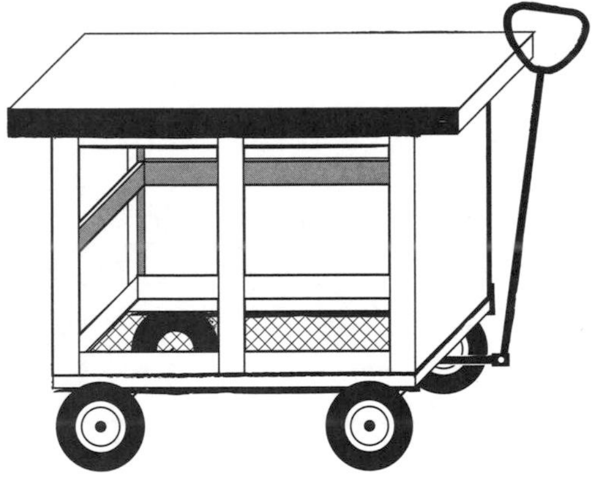

4. Cover the sides and top with FRP panels, leaving the front open for access.

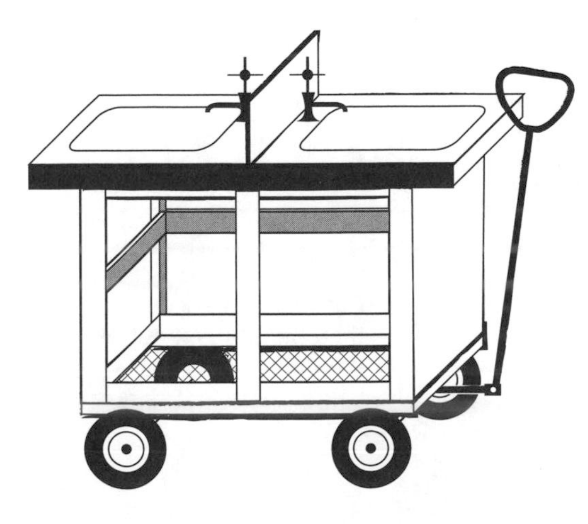

5. Install the sink divider in the center of the sink.

6. Cut holes for the sinks and install.

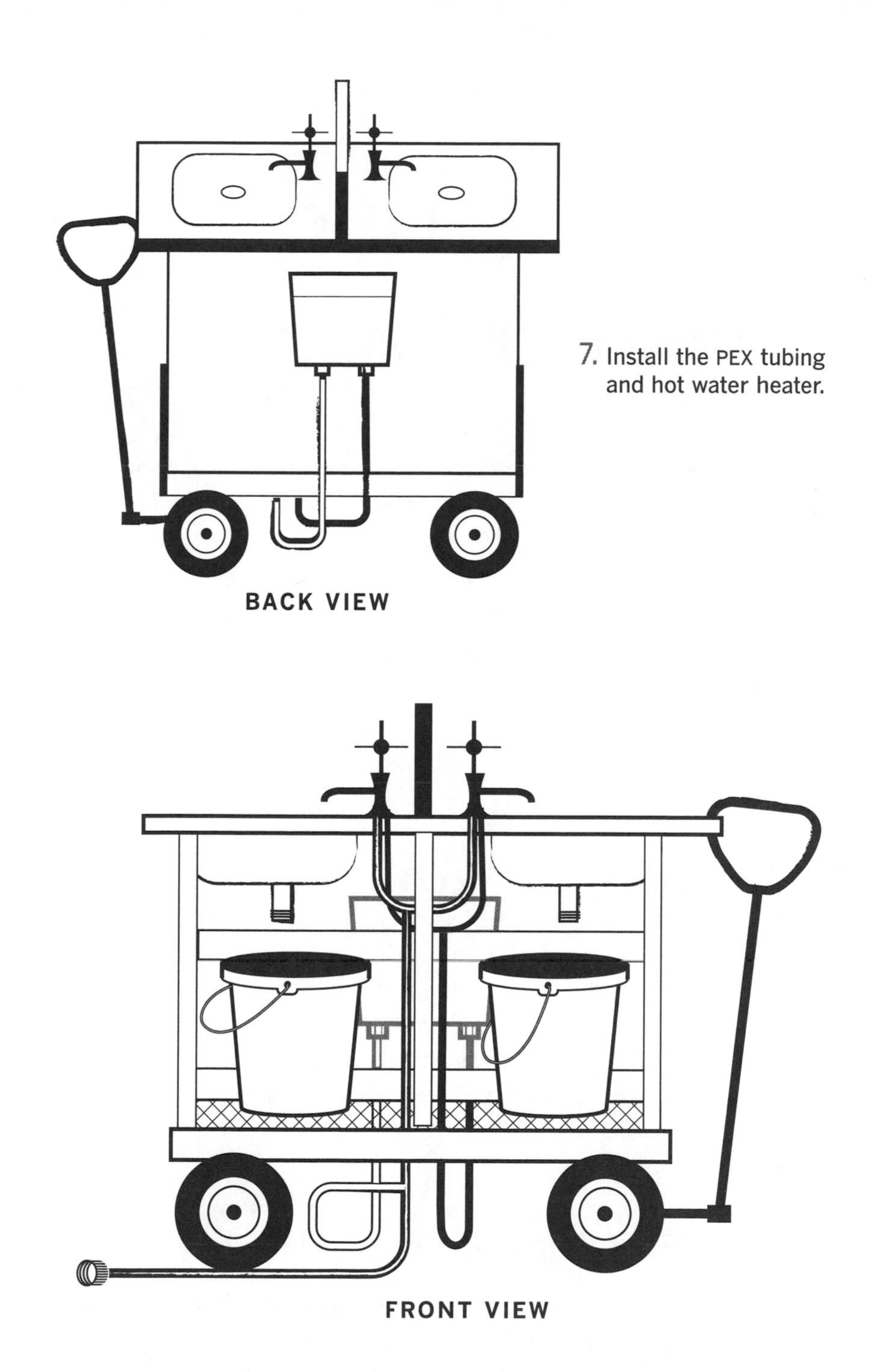

7. Install the PEX tubing and hot water heater.

8. Plug in the heater and wash your hands.

4

Training the Chicken Crew

It's hard for me to explain to people who don't work with chickens how this could possibly happen, how affection for chickens could eventually evolve into slaughtering and eating them. How that could possibly be logical? But if you provide them with quality of life, and you care for them properly, and slaughter them in a humane way, it's an honorable relationship. I'm proud of the work I do.

— Emily Palmer, farmer and
IGI Chicken Crew member, 2010–11

YOU'VE PULLED TOGETHER YOUR CHICKEN CREW. Now you need to train them. It's best to hire a professional to train your nascent crew, whether they have experience in poultry slaughter and processing or not. This will get everyone on the same page right from the beginning, in terms of animal handling and welfare, safe equipment management, understanding regulatory requirements, and food safety.

When you order your poultry processing equipment, the distributor may know of a capable, credible person who can both deliver the MPPT and help you get started.

The Crew's First Training: Live and Local

Have the first training of the Crew on a working farm. This is far more desirable than going to a workshop away from your community or a classroom or conference room. The Crew needs hands-on experience in how to off-load, how to set up the MPPT on a farm, and how to clean the equipment.

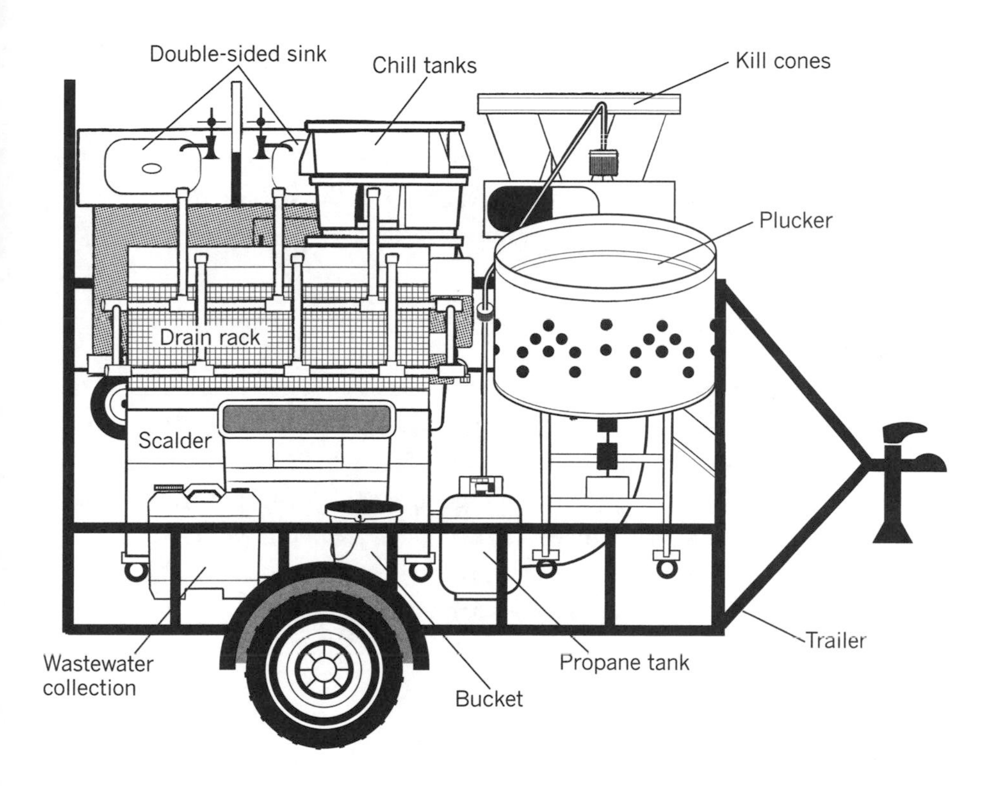

The MPPT*: the whole shebang. Don't forget to check your tie-downs before heading out on the road.*

Situations, questions, and concerns will come up that are unique to your community, your inspectors, and your environment. Training the Crew close to home will help get everything out in the open for farmers, the trainer, the Crew, and the organization managing the MPPT.

The first training must include live birds. The chicken resulting from the training will be considered *custom*: that is, it should not be sold. It is for home use only. So have a feast later.

Invite the farmers in your community so they can meet the trainer and the Crew. (If the Crew is trained far away, these will be missed opportunities.) Training days are terrific occasions for farmers to connect with the Crew and your poultry program. This first training is crucial to the trust and relationship building that is a vital part of the program.

It is important that farmers and Crew learn in a safe and welcoming environment. They must be able to ask all the questions they have and they need to get their hands clean, then dirty, and then clean again. Understanding how the scalder and plucker work, tagging birds, the temperatures required for food safety, how to set up, how to eviscerate a bird: a lot of information is dispensed at a training session about the entire process from crate to bag to refrigeration. All players must be on their game.

Bringing in the Regulators

You may also want to schedule your trainer to meet with boards of health during his or her visit. This is preferable to having the regulators actually present during the inaugural training, which might intimidate and inhibit the Crew and the farmers. Best to set up another training/demo for a future date that includes the agents. Schedule a live-to-dressed bird demo at a neutral location for all the boards of health to observe the Crew in action once they are trained up.

Suggested Board of Health Guide for Chicken Slaughter Inspection

Here's an overview of the board of health expectations your unit must meet or exceed, in roughly chronological order.

1. Birds. Maintain optimal welfare and health, including handling of bird and holding prior to slaughter.

2. Site. Must be clean and uncluttered, with short grass and no overhanging trees; gentle gradient away from clean side and evisceration; food crops located at a sufficient distance and upwind. Weather (wind and temperature) must be calm and moderate.

3. Site setup. Waterline must be clean. Maintain adequate clean-dirty separation. Have all equipment pre-sanitized, keep spare knives in a sanitizer, and confirm sanitizer strength. Provide sufficient wood chips under kill cones, plucker, evisceration table. Keep hand wash stations fully stocked. Store an adequate amount of ice in a protected container.

4. Personnel. Crew must be healthy and clean, wearing hair restraints and no hand or arm jewelry. They must be knowledgeable and adequately supervised.

5. Kill side. Confirm that all birds are dead prior to scalding. Monitor the number of birds escaping the kill cone; monitor the scalder water temperature; oversee proper cleanup prior to leaving the kill side.

6. Evisceration. Thoroughly rinse and sanitize the table if feces or ingesta escape; sanitize any knife that gets contaminated with feces. Thoroughly rinse cleaned bird prior to placement in chill tank.

7. Time/Temperature. Ensure less than 40°F (4.44°C) within 4 hours, and maintain less than 41°F (5.0°C) during packing until refrigerated storage or handover to buyer.

8. Packaging. Make a thorough final inspection. Properly sanitize the rack. Ensure that Crew members bag the birds without touching the inside of the bag with their hands, and without touching the outside of the bag with the table. Check that birds are packed in sanitized totes in ice on top and bottom.

9. Transport. Check that truck bed is cleaned and sanitized if delivering to the buyer, with birds secured properly, temperature monitored prior to transport and upon delivery. Scrub and sanitize mobile equipment.

10. Site cleanup. Collect liquid and solid waste, thoroughly clean equipment, take inventory of anything needed (like soap or paper towels), and alert Crew manager of missing supplies.

11. Compost. Follow department of environmental protection specifications; observe for signs of scavenger activity, dense insect activity, smells, and such.

SERVSAFE

Include a food safety course for your Crew in your ongoing education and training program. ServSafe is one company that provides this training. This solid, proactive food safety information resource is important for the Crew to know and will bolster the confidence of regulators. Topics include Hazards and Sources of Contamination, Employee Health and Personal Hygiene, and Equipment and Sanitation. ServSafe offers classes online and in multiple languages including Spanish, Korean, and Chinese.

For information about ServSafe Food Safety certification classes in your area, see Resources.

Almost Kosher

Humane slaughter and animal welfare are always inherently part of the systems developed around the MPPT. The method of actual kill that the Crew uses is essentially that of the ritual slaughter *shechita.*

Two cuts are made across the chicken's throat. There is no stabbing or gouging. The knives are kept razor sharp and are sharpened throughout the day to maintain the edge and avoid any undue pain to the animal. The neck is slightly extended so the skin is taut. The cuts do not nick bone, nor is the bird decapitated. Not only would decapitation be beyond the scope of necessary action, but there is the practical food safety precaution as well. You don't want to put a headless bird in the scalder.

The knives are kept razor sharp and are sharpened throughout the day to maintain the edge and avoid any undue pain to the animal.

The cuts are made with intention: to bleed-out fast and bring a swift death with minimal pain and stress. The slaughterer holds each bird in his or her arms while making the cuts, then puts the bird into the kill cone for its last death throes. Death is confirmed by a Crew member before the bird is transferred from the cone to the scalder. One at a time.

Ritual slaughter is prescribed by both the Torah and the Koran. The birds killed as described above, however, are not kosher by kosher standards. A strictly kosher kill would require a *shochet*, a person who is licensed and trained both in the kill and in the koshering processes afterward.

Finding local kosher chicken is even more difficult than finding local chicken. The local kosher food movement has greater challenges because of the lack of trained *shochet*, much less access to kosher slaughterhouses. Increasingly, organizations work to make kosher food a local, attainable, and more sustainable part of the food system. In the Boston area, LoKo is a small grassroots nonprofit working with the Jewish community around issues of kosher local food and sustainability (www.lokomeat.com).

Indeed, just like any slaughter, it is only as good as the people who do it. And like any slaughterhouse, it will be only as responsible as the people who manage it.

MPPT Map: What Goes Where

Give yourself at least an hour to set up the components of the MPPT so the work can then proceed calmly, briskly, and efficiently, with no rush. Use this map to guide you in what goes where.

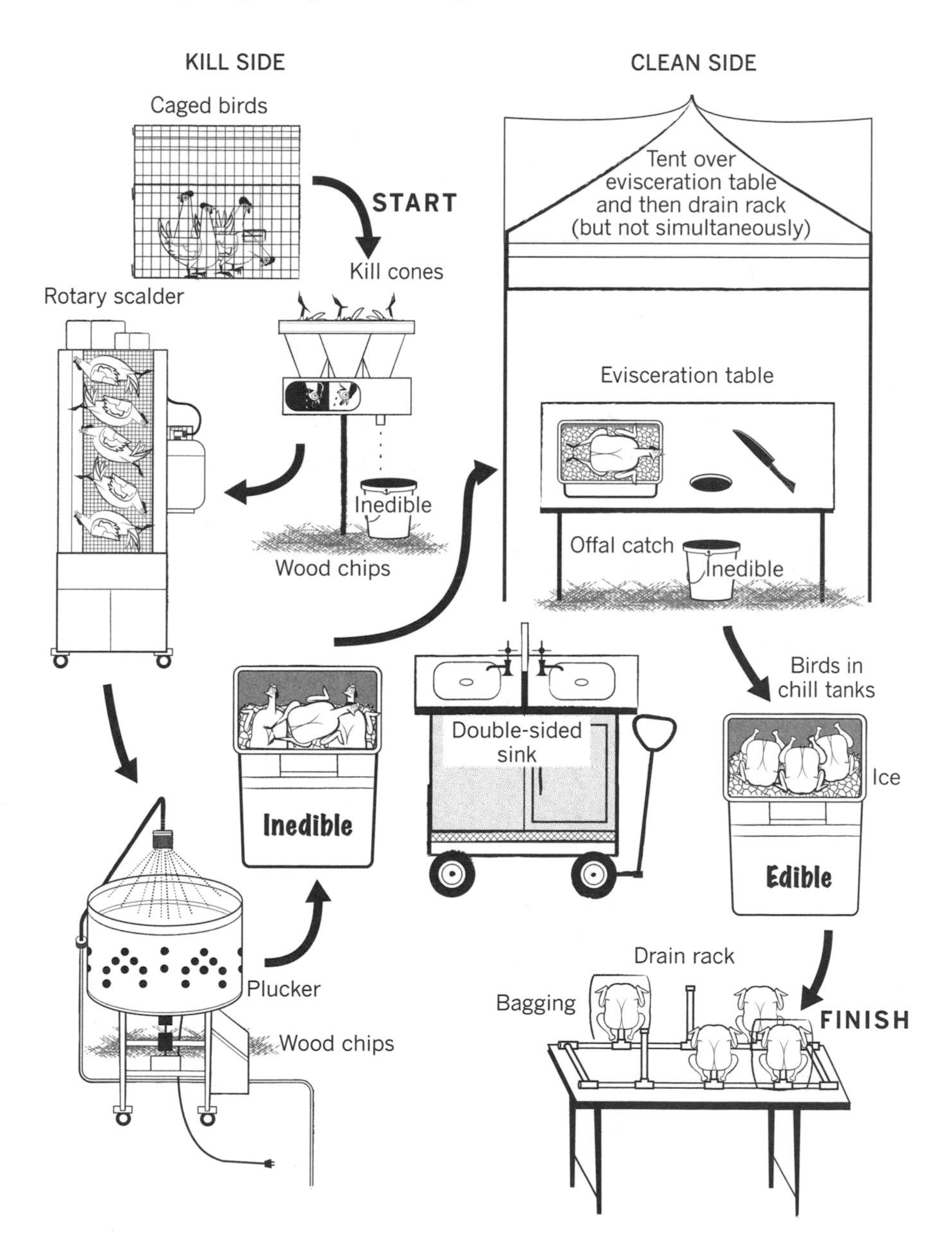

Humane Slaughter, Step by Step

As you begin, the birds should be in their crates, calm, and in the shade.

EQUIPMENT SETUP AND PREP

1. Unload the trailer and place the components in their proper arrangement according to the MPPT map (page 55).
2. Fill the scalder with water and begin to heat it to 150°F (65.5°C).
3. Fill two 30-gallon trash cans with water, two halfway with ice to create an ice slurry, adding ice as needed. Keep the hose available for top-up and refill.
4. Place a 4-inch-high pile of wood chips under the kill cone, plucker, and evisceration table (*not* under the scalder — fire hazard).
5. Sterilize the evisceration table. Sharpen the knives and sterilize them.

HOLDING AND KILLING THE BIRD

1. Hold the bird under your left arm if you are right-handed, and vice versa. Hold it upside-down, its head toward your front, its feet up and back. With the hand that is holding the bird, pull the head down. Wait till the bird is calm.

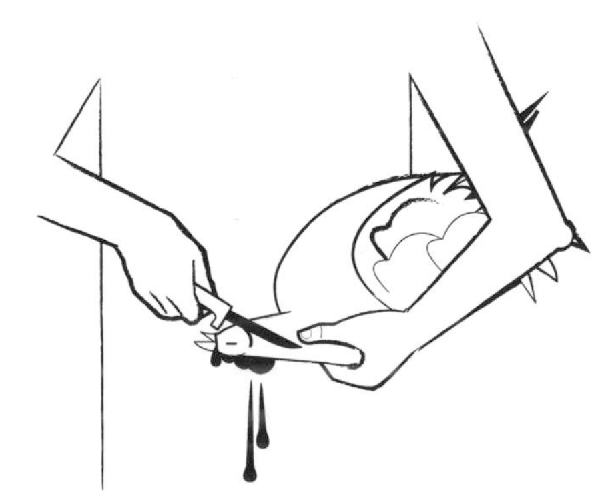

2. With your other hand, find both of the bird's mandibles (jawbones). With a sharp, sanitized knife, cut cleanly, firmly, and quickly across the blood vessel above each mandible.

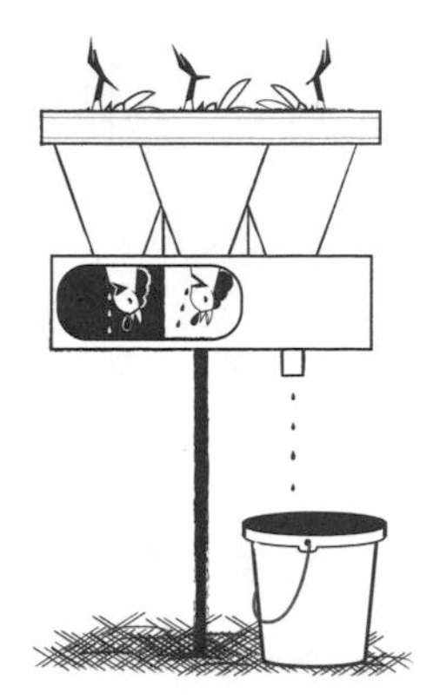

3. Place each bird upside-down in the killing cone while its blood drains.

SCALDING AND PLUCKING

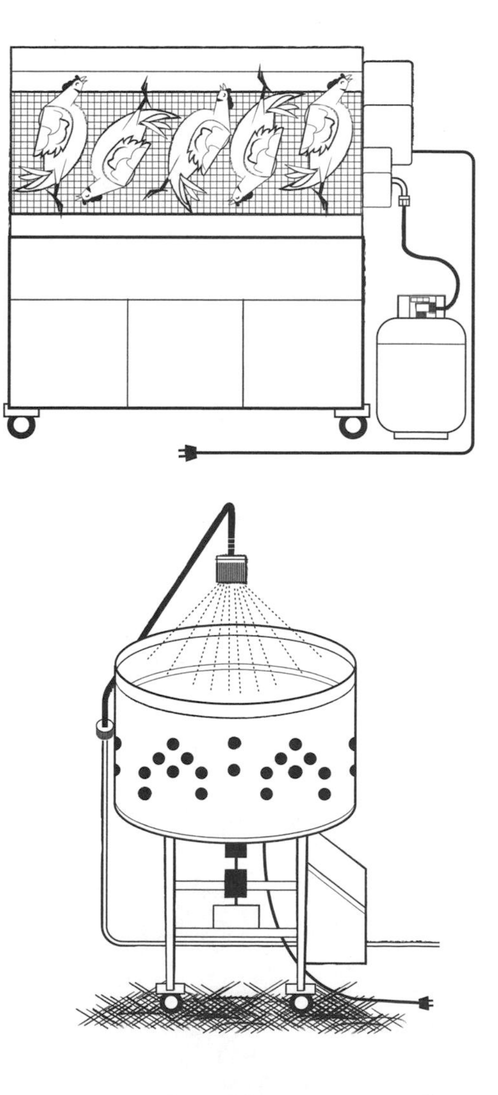

1. Confirm that the water temperature in the scalder is at 150°F (65.5°C) and then slip the bird into the water.

2. Agitate birds to ensure water penetrates the feathers and loosens them at the root. A good scald ensures a good pluck.

3. Test readiness by pulling on feathers. When they come off easily, remove the bird from the scalder. It usually takes 45 to 60 seconds.

4. Place the bird in the plucker and spin it until all or most of the feathers are off. If they don't fall off easily, check the time and temperature of the scalder.

5. Remove the bird from the plucker and place it in the cold water bath till you are ready to proceed with evisceration.

EVISCERATION

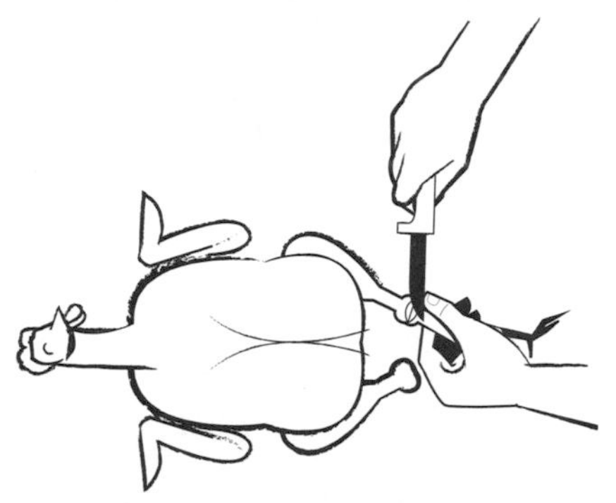

1. With a sharp, sanitized knife, bend the foot down, quickly sever the knee joint, and remove the lower part of the leg.

2. Cut off the head, loosen the neck skin, and separate the crop and the esophagus from the skin.

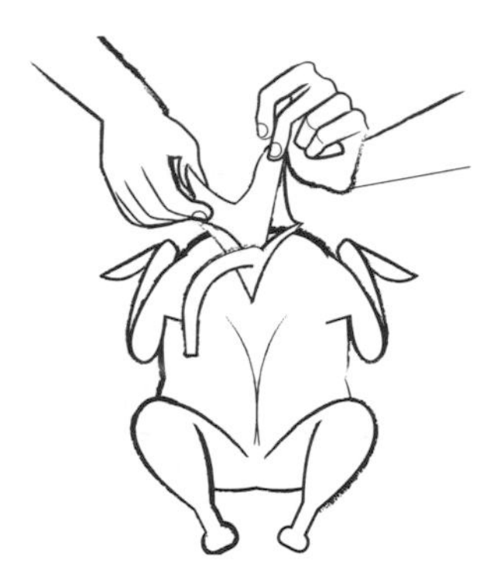

3. Loosen the trachea.

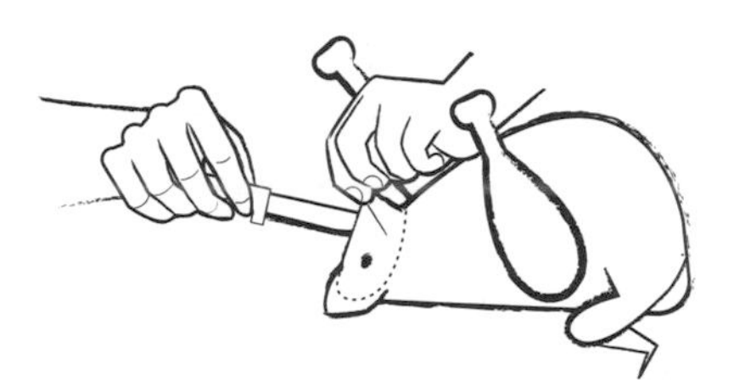

4. Cut a shallow triangle around the cloaca (vent) to create an opening.

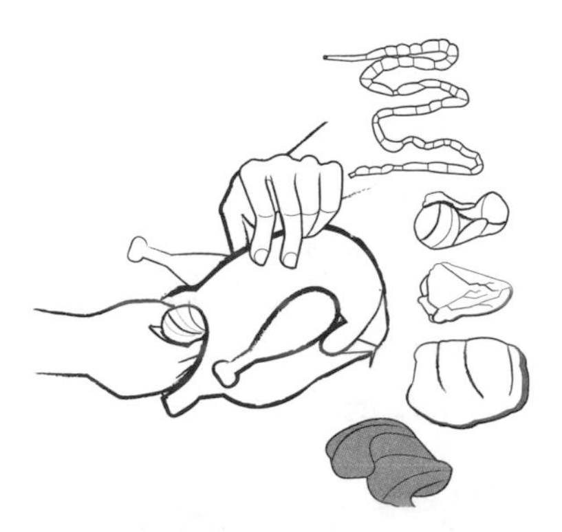

5. Reach in and pull out the organs.

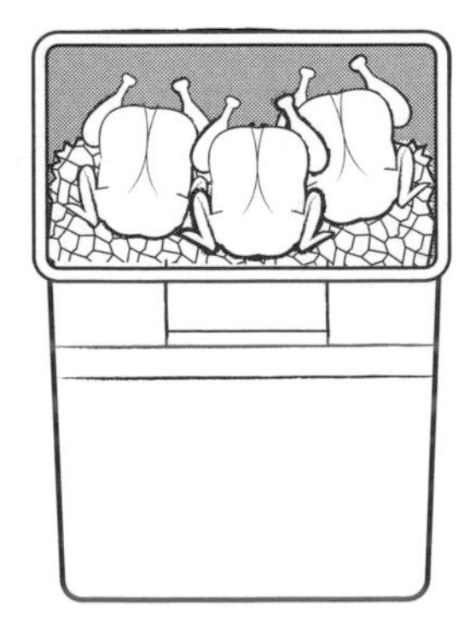

6. Place the chicken in a chill tank awaiting the drain rack.

How to Package a Fresh Chicken

by Jim McLaughlin of Cornerstone Farm Ventures

This is the way to give your packaged chickens a sleek, fresh appearance that inspires confidence in your customer.

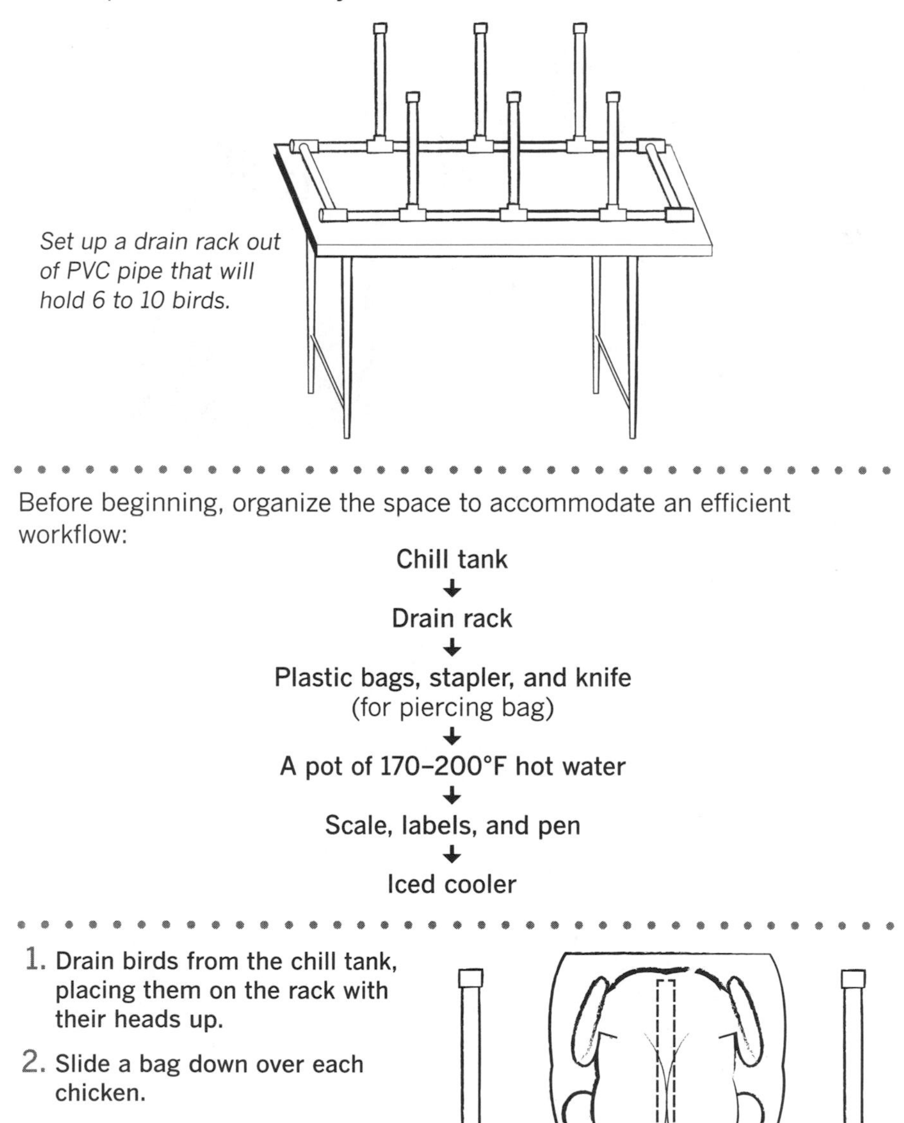

Set up a drain rack out of PVC pipe that will hold 6 to 10 birds.

Before beginning, organize the space to accommodate an efficient workflow:

Chill tank
↓
Drain rack
↓
Plastic bags, stapler, and knife
(for piercing bag)
↓
A pot of 170–200°F hot water
↓
Scale, labels, and pen
↓
Iced cooler

1. **Drain birds from the chill tank, placing them on the rack with their heads up.**
2. **Slide a bag down over each chicken.**

(continued next page)

3. Take one bird off the rack, turn it upside down, and push it all the way into the bag.
4. Squeeze the legs together and gather the bag tightly around the legs. Then spin it to form a "pigtail" out of the excess bag material.
5. Staple the bag. The staple should go *around* the pigtail and not through it.
6. Pierce the bag at the breast area (over the breastbone) or at the base of the clipped end of the bag by the legs, to let the air out.

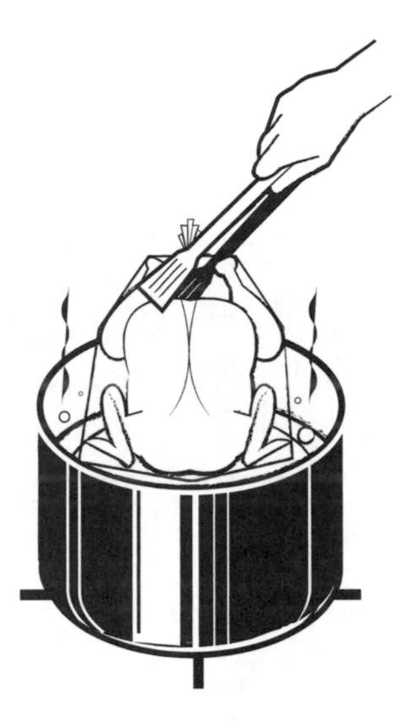

7. Place (dip) the bag into 170–200°F (76.7–93.3°C) hot water. Swirl the bag in the water or use a pair of tongs to hold the bird down (it may want to float). There should be a rush of air and bubbles, so be careful not to get burned. The shrinking is done in an instant. Dip the bag only as long as it takes to shrink (1 or 2 seconds). Do not wait until air bubbles stop as water will enter into the bag.
8. Pull the bagged bird out of the water. Weigh the bird and put a sticky label right over where the bag was pierced at the breast. If you plan to put a label on the package do it now, after wiping off excess water. No need to close the hole if you pierced it at the feet end. Note: If you have a bubble of air inside the bag, it means the bag was not pierced sufficiently to let out all the air.

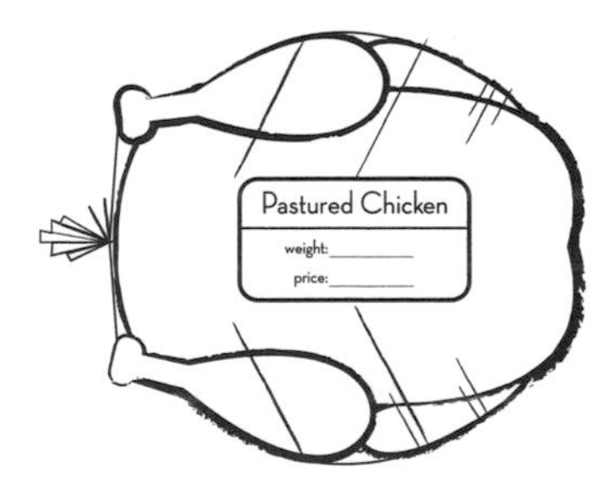

9. Immediately place the package in an iced cooler. Do not place the bird back into ice water. If you do, water will leak into the bag, and labels will not stay on in water.

WHY WE KILL THE WAY WE DO

There are reasons for the methodology — a system of ethics behind each step.

WHY WE DON'T KILL THESE WAYS

Using an ax. It's dangerous, it lowers food quality, and a decapitated carcass in the scalder is gross.

Breaking the neck, spinning, or piercing/stabbing the bird through the roof of the mouth or throat. These methods torture the animal.

Gassing, stunning. These are industrial/mechanical solutions that, in my view, remove any shred of humanity.

THE WAY WE DO IT

We kill this way because it gives us and the animal pause. Because this way reveres life and minimizes pain.

You will hold the chicken securely and calmly in your arms and gently be sure the skin at the neck is taut. The whole idea of the cut(s) is to lower the blood pressure to the brain while the heart pumps to bleed out. One to two quick cuts to the carotid arteries with a razor-sharp knife will do it.

Close-up: Tips on Holding and Killing a Bird

by Jefferson Munroe

The birds should be crated (though not overcrowded) near the kill cones. When you pick up a chicken for the kill, try to keep the entire process as quiet and subdued as possible. Any commotion probably means that something is going wrong — there is an increased possibility for broken wings or birds not dying as quickly as one would like.

The following is for right-handed people, so reverse if need be. Be sure your knife is razor sharp. I test the blade on the hair on my left hand.

PICKING UP AND HOLDING THE BIRD

1. Open the bird crate for as short a period of time as possible to avoid flapping and escaped chickens.
2. Grab the bird by its sides and gently pick it up.
3. Once the bird is in your arms, turn it upside down — this rushes blood to its head and generally calms it.
4. With your left hand pin the bird with its back against your chest, and firmly grasp the chicken's ankles with your right hand.
5. Next, move your left hand to the neck of the bird and, still cradling the back of the bird against your chest, use your right hand to tuck the feet into your left armpit.
6. When you let go of the feet you will have a chicken cradled in your left arm, leaving your right arm free to wield a knife.

MAKING THE CUT

1. At this point, the key to a quick death is ensuring that the skin around the carotid arteries (directly below the jaw line) is taut. I take the thumb and forefinger of my left hand and place them just below the base of the jaw line.
2. From this point I draw the skin back behind the head of the chicken and pinch it together the way one would hold the scruff of a dog or cat. Pulling the skin this tight will constrict the airflow of the chicken, so I try not to hold it for too long. If you're practicing you can pull the skin taut and then let go a few times to make sure you've got it right.

THE FINE POINTS

When you're ready to make the cut you must slide your knife rather than saw it — your knife is very sharp, but you need to use the edge. This point in the process is one of the only times a very sharp knife is very close to fingers, so be sure you're holding the bird tightly.

3. With your finger and thumb behind the chicken's head and the skin taut below the jawbone or mandible, make your cut ¼" below the jawbone. I try to follow the jaw line.
4. When you make the cut, the blood should spring out of the neck for a second. If it dribbles out then you probably haven't cut deeply enough and the bird will take longer to die.

5. I make a second cut to the other side to ensure that death comes quickly, but one cut should be enough if the first cut is properly done.
6. When you make the cut the bird should not flinch — if it does your knife isn't sharp enough. Blood will start flowing and the bird's eyes should flutter as blood pressure to the brain drops and consciousness begins to leave it.
7. Let go of the skin with your left hand. There should be a constant stream of blood from the bird's neck. If not, make another cut or two to ensure a quick death.
8. Once you are sure the bird is bleeding properly move it into the kill cone quickly. If not the wings will start flapping and there is a higher chance of broken wings or you getting slapped in the face by a wing.

Congratulations, you just killed a chicken, humanely.

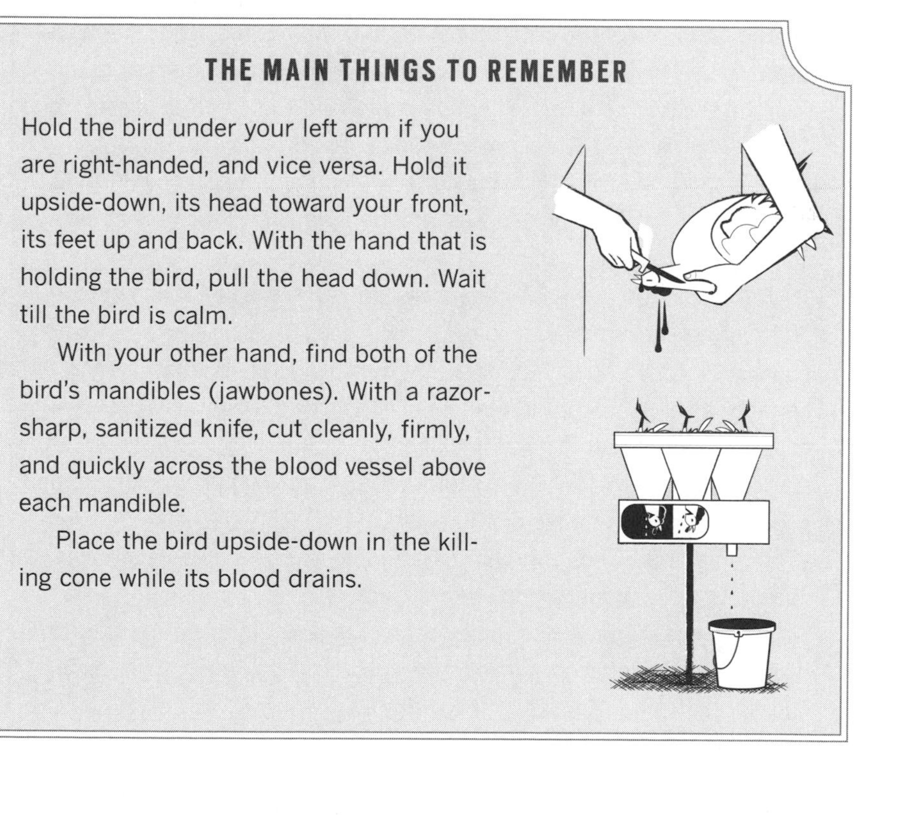

THE MAIN THINGS TO REMEMBER

Hold the bird under your left arm if you are right-handed, and vice versa. Hold it upside-down, its head toward your front, its feet up and back. With the hand that is holding the bird, pull the head down. Wait till the bird is calm.

With your other hand, find both of the bird's mandibles (jawbones). With a razor-sharp, sanitized knife, cut cleanly, firmly, and quickly across the blood vessel above each mandible.

Place the bird upside-down in the killing cone while its blood drains.

Blood and Humanity

You recognize the smell of blood from the inside out. At this slaughterhouse in South Dakota, wearing haz-mat coveralls, rubber boots, hairnets, and hard hats, we all looked the same — Temple Grandin, myself, the man operating the hydraulic squeeze chute that holds the cattle, the rabbis, the people in chain mail wielding saws and sharp hooks working the processing floor — allied in this weird hallowed hall where animals walk in and meat comes out. The hum of a slaughterhouse is the smell of blood and disinfectant.

IN THE FIELD WITH COWS AND TEMPLE GRANDIN

I was lying in a field in South Dakota, looking up. It was spring beautiful, the end of a long few days. The pasture was green gone. The wide-open vesper of the South Dakota plains merged possibilities of beauty with the realities of dirt, sun, rain, fertility, and the smell of clover, trodden grasses, and a herd of cows, heavy with life.

"Lie still," she said, "and the cows will come up to smell your hair." She chuckled as she spoke. I was with Temple Grandin. If she could laugh lying so close to so many cows, so could I.

Temple Grandin is a livestock behaviorist. She has Asperger's syndrome. She is a rock star, and she is an angel. These facts and metaphors are not in opposition. In reality they are connected at the core in perfect harmony. For Temple relates to the world as animals do. She possesses amazing gifts and intelligences that have forged the highest standards of livestock welfare around the globe. And like rock stars and angels, she has her share of detractors.

Temple and I were enjoying the end a long day at the slaughter and processing plant for the Dakota Organic Beef Company in Howard, South Dakota. Temple was there to consult, and the owners hoped to receive her seal of approval and humane certification for their kosher and non-kosher systems. For a few days as a visitor I was awestruck, privileged to watch Temple work, happy to drive around with her in a pickup truck eating frozen Snickers bars and listening to classic rock on the radio.

A slaughterhouse must be calm. There should be no scent of fear in the air. If I smelled it or the animals smelled it, it emanated from me. The only possible trepidation I picked up was on the faces of the workers, not the animals. Why was this stranger in their midst? Who would want to visit a place such as this? Fear transmits, and we humans are as much a herding animal as cattle are. Easing into calm and peace with this place was the only way to cope. The bright lights, the hydraulics and saws, the blasts of hot water, the carnal smells, the workers in hard hats and rubber coats, their aprons stained with flesh, blood and bone: all this and yet it *was* calm. It worked. I breathed it all in and in return learned great lessons from Temple, from the rabbis, and from the processors who worked together in that place.

Here humane slaughter took on meaning beyond theory or rote definition. The ritual slaughter of a kosher kill leaves no doubt that the stroke of blade against throat means blood and death. In contrast, the action of a stun gun to the head of an animal means collapse, hoist, and then blood flow. In this scenario, the responsibility of the kill lies on neither the operator nor the hoister. It's like a firing line: no one, then everyone, then no one is responsible.

I breathed it all in and in return learned great lessons from Temple, from the rabbis, and from the processors who worked together in that place.

By blade, though, there is most definitely a killer, a slaughterer's hand that fells a fellow creature. There's no denying the connection between life and death when there is blood in those crucial and important moments. Once those moments are mechanized and obfuscated from view, they make one more link in the chain of disconnects, giving us permission to be complicit in the treatment of any animal as another cog in the machine. Without blood, and by justifying its absence, we remove ourselves even further from the relationship of man to creature. We do know better and we do remember the difference, but we just choose not to.

In ritual slaughter, the responsibility lies squarely on the shoulders of one person. The act must be done with care, training, and the intention to act for humanity. In the end, humane slaughter keeps us human as well.

5

Education, Marketing, and Outreach

> *A peasant becomes fond of his pig and is glad to salt away its pork. What is significant, and is so difficult for the urban stranger to understand, it that the two statements in that sentence are connected by an* and *and not by a* but.
>
> — John Berger, "Why Look at Animals?" in *About Looking*

THE MOBILE POULTRY PROCESSING TRAILER (MPPT) described in the preceding pages is the cornerstone of a poultry program. But it will remain a pile of dysfunctional stainless steel in the back of a trailer, rotting away in the back pasture or someone's barn, unless it is supported by three important driving forces:

- Community
- Advocacy
- Outreach and education

Education

It's all about education. You must educate both your Crew and your community, and this includes your regulators and our policy makers.

Community Poultry Day Workshops

Poultry Days are such fun. You can keep them low-key and low-cost, and they'll still make big impacts. They are relaxed opportunities to inform the growers in your community about the MPPT, to introduce the Chicken Crew, and to share information that helps people raise healthy, happy birds and happy healthy chicken dinners. They are a chance for farmers and backyard growers to share information in a comfortable, safe setting.

Hold a Poultry Day in the off season (not your broiler season). In New England, we hold ours in February. It is a great excuse to come together in the winter, share information, and get excited about the upcoming season. Farmers connect with other farmers. Backyard growers meet like-minded folks and ask questions in a supportive setting, even if they feel their 10-bird flock is too minuscule to bother anyone about. Mentoring relationships between the experienced and the novice develop comfortably.

Most important, Poultry Day is the chance to tell anyone and everyone who might be interested in using the services of the MPPT and Chicken Crew, to schedule their slaughter and processing dates the day their chicks arrive in the mail.

Funding Your Poultry Day

Make your event free or ask for a suggested donation at the door. Perhaps a local feed store or restaurant will help sponsor your Poultry Day or donate to defray any costs. Suggest that they support this education outreach as an investment in their community while the poultry program develops. This is short-term money for a long-term benefit.

Consider making and selling T-shirts that promote your program.

Poultry Day Suggestions

Hold the event in a public space, accessible to everyone, where you can serve food — a Grange Hall, school, or community building, for example.

Start simply. Schedule a straight-up day of events and send out the schedule so people can plan to attend the workshops that interest them.

Publicize your Poultry Day. Write a press release and send it to local papers two to three weeks before your event.

POULTRY DAY CHECKLIST

- ☐ Information
- ☐ Refreshments
- ☐ Handouts:
  - ☐ Chicken Crew business cards or contact information to give to growers
  - ☐ Farmer's Checklist for the Day Of (see page 127)
  - ☐ Equipment catalogs and chick catalogs from local feed stores
  - ☐ Materials from your USDA Rural Development office and Extension service, regarding their programs
- ☐ Slideshow on computer, such as chicken tractors, roast chicken, happy farmers, happy birds. Ensure in advance that projector is compatible with computer.
- ☐ Screen or sheet to project images onto
- ☐ T-shirts or other promotional products you develop to raise money

Contact your local radio stations. Commercial radio stations have community calendars as well as spots for free public service announcements; local stations may interview you or set up a call-in show. Local NPR stations frequently have community calendars where you post your workshop information online for free.

Be ready to give interviews. Media outlets may want to talk to you before the event — or come out to get some interviews, hear some chickens clucking, and try some coq au vin or chicken stew.

TALKING POINTS

Here are some typical questions and possible answers.

Q: *Why is your poultry workshop important?*

A: It connects growers, provides an educational opportunity, builds mentoring relationships, shares information, and supports your local food web.

Q: *Who are you trying to reach?*

A: Farmers, backyard growers, eaters!

Most important, be sure to say when and where the event is. And thank your donors, speakers, participants, and the farmers who raise chickens.

Mix up your speakers. Invite local experts and one or two from neighboring communities or your state's agriculture department. Or show a relevant documentary film. According to a friend who has run an annual Farm Film Fest, films are surprisingly affordable and the directors are eager to share them.

Set up a resource table. Include books, calendars, cookbooks, feed information, and pamphlets from your state's department of agriculture. Include a range of information that appeals to different levels and age groups. Kids can raise chickens too!

Film your Poultry Day for community television if it exists in your community. In our experience, these segments get a lot of airtime.

Make It Delicious! Make It Delicious! Make It Delicious!

Poultry Day is always a big hit when the day's events close with a workshop titled "Cooking with Local Chicken." (Don't fake it here — be sure it's really local!) Because small-scale poultry production usually translates into a higher price per pound, and eaters will most likely end up with a whole bird in their hands instead of cut-up parts, help your Poultry Day participants learn how to cook this wonderful-tasting bird economically. (See recipes, pages 110-125).

TIPS FOR A SUCCESSFUL WORKSHOP

- Stay on schedule.
- Invite your local regulators, state regulators, commissioner of agriculture, local representatives, Extension agents, congressmen, and senators. A Poultry Day is a terrific opportunity to be transparent and forthright in your community.
- End your day with a community potluck and the sharing of chicken. In our community, the Martha's Vineyard Agricultural Society dovetailed into the Poultry Day by hosting a community potluck in the same evening. It was a wonderful time to come together in the dead of winter to gather and talk farming and cooking.

Workshop Topics

Consider mixing and matching your Poultry Day's workshops from three subject areas: General Chicken, MPPT-Specific, and Home Cooking with Local Chicken.

GENERAL CHICKEN WORKSHOP

Here's a sampling of topics addressing general chicken care.

- Choosing a Breed (Heritage or Cornish Rock Cross?)
- Predation Prevention
- Mobile Pasturing Pens (chicken tractors): plans, different styles, how to build, how to use reclaimed or recycled materials
- Brooding your Chicks
- Weather Extremes: Protecting your Flock
- Marketing
- Using Social Networks and Social Media
- Composting Inedibles

MPPT-SPECIFIC WORKSHOP

These topics cover aspects of the MPPT that interested farmers might want to know about, but they do not substitute for actual training (see chapter 4).

- How to Prepare Your Farm for the MPPT
- Overview of Regulatory Requirements a Farmer Must Meet to Sell His or Her Birds
- Meet and Greet the Chicken Crew
- Scheduling the Chicken Crew

"HOME COOKING WITH LOCAL CHICKEN" WORKSHOP

Connect with a local chef, private chef, or home cook who is comfortable with public speaking while cooking. This individual should be able to share information and tips on how working with a local, whole chicken is different from cooking with a commodity chicken.

Decide with your presenter how best to show the differences. Here's one scenario:

- Have the following types of chicken on hand: a stew hen, a rooster, a commodity bird such as a Tyson or Perdue chicken, a heritage breed bird, a pastured cross.
- Talk about the differences among the types.
- Decipher the labels: natural, organic, free range, etc. (See pages 109–110 for more on label terminology).

Marketing

Selling a chicken, or anything you grow, is an opportunity to talk about how it was raised and how it was slaughtered. You're given this moment to inform your customers, so they can learn what it took to get that bird from the field into their pot — from feeding the birds to protecting them from raccoons and hawks. From the days it was so hot you might have lost some to weather, to how and where they were slaughtered: overall, you describe what it feels like to raise an animal and then kill and eat it.

It's also a chance to share with them what farming means to you, why you do what you do, what you value about it. And your eaters, let's hope they vote, because if you need their help to mediate with zoning boards, or regulators, they'll be by your side.

You have to be prepared, though, and lay the groundwork.

Develop a Spokesperson

Identify a person who can communicate the goals of your poultry program to your community and to the media. This spokesperson needs to know and be able to explain the reason for your program and its impact in succinct sentences that have a beginning, middle, and a period at the end. They should be consistently available to answer queries and write and send press releases.

Get to Know the Media

Knowing the media outlets and their deadlines will help when you need to announce events, fundraisers, and milestones. Even a seemingly benign event, such as the new availability of local chicken at a neighborhood restaurant or greengrocer, deserves recognition. A press release that is accompanied by a photo will usually get better placement and always be more effective.

You have the least control in the print media. Editors on deadline tend to cut from the bottom up. With radio, television, and video you're not as vulnerable, but it is crucial to be clear, comfortable, and on point.

Get to know the food writers, bloggers, and journalists who cover areas such as agriculture, policy, food, and cooking. Tell them your story, and keep them posted as the chapters unfold.

And once you are awarded a USDA grant or loan, don your uniform (such as a pink *Local Meat* T-shirt), get out your camera, and snap those hand-shaking, contract-signing, check-accepting photos for your website and beyond. Manage all the while the message fundamentals of your program: safe, affordable, accessible, permitted, clean, size-appropriate, humane slaughter and processing of poultry and good food to feed your families.

Straight Talking

When you're talking to customers or the media, use the language of slaughter and processing straight up. Vegetables are harvested; livestock are slaughtered. Processing is the transformative act that takes place after an edible animal has been slaughtered and turns it into a safe food. Butchery is the act that breaks down the raw meat into parts such as thighs, drumsticks, and breasts.

Animal welfare should be upheld at all times and emphasized when you speak, write, and communicate information about your poultry program and the MPPT. Humane slaughter is not an oxymoron. It is a mandate, an obligation. Inhumane slaughter is reckless and unacceptable.

GOOD GRAPHIC DESIGN

Labels, posters, recipe cards, information about your poultry program, anything that involves ink on paper — design it well. Good design and readable fonts make a huge difference. If your materials are difficult to read, illegible, hard to follow, or don't suit the purpose or message you're aiming for, you will lose your audience, and the important information you're trying to convey will fade to black.

In this case, a look that is clean, clear, simple, and honest will eloquently communicate, in visual terms, the values of the poultry program and the MPPT. Good design is good communication.

Outreach

There are all kinds of ways to excite and ignite your community. For example, it's important to highlight all the facts and figures on your side (see box, page 73). Impress your family, friends, and neighbors!

IGI'S MPPT: FACTS AND FIGURES FOR YEAR ONE, LICENSED

IGI processed, on average, 80 birds per process, and the MPPT was used about two days a week in its first licensed year.

- Licensed Producers (under the first year of license #417) and their number of processings:
  - The GOOD Farm: 21
  - Morning Glory Farm: 8
  - Cleveland Farm: 4
  - The Whiting Farm: 3
  - North Tabor Farm: 4
  - Flat Point Farm: 1
- Number of birds processed under IGI's permit: 3,580
- Estimated increase in grower revenue: $72,000
- Approximate new farm wages created: $18,000
- Estimated increase in grower profit: $39,000
- Number of backyard growers: 13
- Estimated number of backyard birds processed: 675
- Estimated total number of birds processed: 3,580 licensed + 675 unlicensed (home use only) = 4,255

Who Knew There Was a Chicken Season?

Since we've devolved into a culture of disconnect from animal to meat and a society of instant-gratification, it may be surprising to an eater to learn that there is such a thing as a chicken season — just like the strawberry season, cherry season, artichoke season, and asparagus season. In our regional food system in New England, our fresh chicken season runs from about April to November. Because the farmers sell out of birds, if I don't get a couple in my freezer by October or November there won't be a chance to buy local chicken till spring.

This is one more way to educate your market — your eater. Let people know that there's a season for chicken, depending on what part of the country you live in, how small your operation is, and how it fits into your overall growing scene. This is a terrific chance to return authenticity and truth to marketing. Use it to your advantage over the "big guys."

Building Community Understanding

Start a Sustainable Book Club. Infiltrate! If you're in a book group already, suggest a title that explores industrial agriculture or a related topic, such as nutrition, food politics, history, or animal welfare. By opening up the dialogue about the poultry industry, you will garner support for chicken in your local or regional food chain.

If you're not in a book club, encourage your local library to select a food-related topic. Or start a club! (Lead: www.oprah.com.)

For a list of various titles and topics specific to factory farming, animal welfare, food politics, cooking, butchering, and more, go to the book list in Resources.

Bring an inspiring speaker to your community. Use the forum to help educate the public about why local food and especially local meat are important.

Give your speaker a tour of farms in your area. Have him or her meet separately with farmers as well. When IGI hosted Joel Salatin to speak one cold November night, more than 300 people showed up. The next day he met with about 30 farmers so they could talk in a more intimate setting.

WHAT'S A CHICKEN TRACTOR — AND WHY?

A Chicken Tractor is a mobile coop, one you move regularly. A chicken tractor provides access to grass, bugs, stones, and weeds for the birds and offers shelter and safety from predators and the elements. A chicken tractor adds fertility to the soil. The chickens in the structure are integral to the tractor: as they scratch, beat down grasses, peck, and poop, they work the land under their movable structure. A chicken tractor is all of it: animals, wood, and wire.

Hold a film screening. Connect with a local independent cinema or film festival or host a house viewing. Or gather your group and watch one of the many talks recorded on video. TED Talks (short for Technology–Entertainment–Design; see www.ted.com) presents important, timely ideas in conferences and provides them for free in video form to the world. TEDxManhattan's "Changing the Way We Eat" is an independently organized event, in this case focusing on sustainable food and farming. It has a website and a YouTube channel.

Create an e-mail newsletter. Some tips:

- Write a clear and interesting subject line. For example, reiterate the name of your group and identify the theme of this edition of your newsletter: "CISA: Backyard Chickens and Zoning." Or "Homegrown.org: Cook and Preserve."
- Or perhaps you've an action alert: "Farm-to-Consumer Legal Defense Fund: Fight GMOs & Support Local Meat Production in the Farm Bill."
- Include short informative news and narratives about the topics and events you are promoting. Keep the entries concise by providing a link for further reading.
- Include a few images, but not too many. If you have more pictures to share, include a link to your website or a photo gallery or share them in an album on Facebook.
- If you have many entries, add a Table of Contents at the top of your newsletter for easier navigation.
- Be consistent in your delivery schedule. Whether you're doing it once a month, every two weeks, or four times a year, keep to it, as people will come to expect it.
- Archive your newsletters on your website.
- Track your stats via your e-mail newsletter provider. Are people opening your e-mails? What time of day are you sending them and do you notice a higher or lower success rate accordingly?

Social Media: Mobile Technologies and Why Use Them

A cursory introduction into the world of social networking:

The Rolling Stones are on Twitter. Mick Jagger, Keef (that's Keith Richards), Ronnie Wood: each has his official "@" except #charlieistoo coolfortwitter. The Boss tweets @springsteen, all lower case.

The Dalai Lama tweets. He has millions of followers and follows no one, as one would expect. The USDA is on Twitter in Spanish as well as English. Animal Welfare Approved tweets. Joel Salatin is a righteous tweeter. Wendell Berry may never tweet, but people tweet about him all the time. KFC, McDonald's, and Taco Bell all tweet. Michael Pollan and Marion Nestle tweet; so does American Farm Bureau, and they probably follow both Mr. Pollan and Ms. Nestle to see what they're up to next.

Whether you're into it or not, social media is part of the media landscape. Best to learn about it and use it for good, not for evil. If you're a farmer, advocate, or restaurant owner, social media can help you manage your message, get the word out about what you are doing, educate your audience, and increase awareness; and it can also backfire on you if you don't use it well.

Today it's all about Facebook, Twitter, YouTube, Instagram, Tumblr, Pinterest. These will run through their paces and in time others will emerge. Still, it's unlikely that social media will ever go away, so it's best to embrace them and use them to your benefit. Think of social media as another tool in your tool kit. Use these tools to your advantage. They're usually free (no fees to sign up) but they do take time to manage.

Of course, the nature of the World Wide Web also means that anything you write, photograph, and post on Facebook or tweet into cyberspace will never go away, either. Consider that before you push Post or Share or Comment. Tip: Monitor your Facebook page for scams and spam (unwanted solicitations and/or inappropriate posts). Also monitor online discussions, to maintain respectful dialogue.

Reporters and journalists, calendar editors and community groups troll for information. Whenever you post a Poultry Day Workshop on a sustainable food group's calendar, or list a fundraising dinner or a MPPT training, you are inviting media attention.

People love "behind the scenes" stories about farms. Share these with your local community through Facebook pages, Twitter accounts, and Instagram. Just be sure to read these links, news stories, and articles before you decide to post or retweet them.

FACEBOOK

What started as a social network for college students has turned into an incredible communication tool for all types of organizations: libraries, schools, nonprofits, publications, and much more. Facebook is excellent because it crosses over into so many communities. It can help start a revolution!

On Facebook there are personal pages and there are fan pages. Be aware of this when you sign up. Keep your personal page for sharing photos with friends and family about your summer vacation and kids' first day of school. Use your fan page to get out the word about your business or your nonprofit.

Post information on your Facebook fan page that is relative to your farm product — recipes, news about recalls, photos, farming grants, consultants, speakers, farmers' markets, catalogs you like, and more. Also use Facebook to help promote educational opportunities for your constituency, such as upcoming webinars, conferences, or radio programs.

EXAMPLES

Here are some ways farmers, restaurants, food advocates, and eaters use social media to make their case.

FACEBOOK POSTS

In these examples, restaurants are promoting the local farm products they were using that day:

> "Lunch Special: The GOOD Farm chicken liver, butternut squash, local pea shoots and bacon-onion jam on sourdough!!!!"
>
> Posted by 7a–Martha's Vineyard, a restaurant in West Tisbury, Massachusetts

> "One Cow T-Bone Island Steak and Eggs over Island fingerlings & kale $15.95 Thanks Pilot Hill, Stannard & Blackwater Farms"
>
> Posted by Scottish Bakehouse (see page 125)

TWEETS

Here are examples of tweets between a farmer and a restaurant:

> **@follownathan** Nathan A. Winters, a farmer in southern Vermont, tweets, "Hey y'all if you are looking to #FF a classy establishment that serves #realfood in #SoVT check out @thewilmington #VT #localfood"

He mentions The Wilmington, a restaurant that supports local farmers, which then tweets from its account:

> **@thewilmington** "Picked up eggs for breakfast from Look and Lundsted farm. #realfood #ag #foodies #wilmingtonvt #SoVT pic.twitter.com/IDHOuBu1"

The restaurant smartly attached a digital image of roaming hens.

TWITTER

Like Facebook, Twitter is free. Sign up here: www.twitter.com, and watch a while before you jump in. You have only 140 characters (including spaces) to say what you want. As always, be careful out there. Post and tweet well and mindfully, as the Dalai Lama does. Like a linguist, learn the language of Twitter and how to use it most effectively.

Unlike Facebook, you need not be accepted as a "friend" by another. I find it useful for learning about news, articles, and commentaries posted by people I follow who are also working with food systems, slaughter, meat, and policy. Twitter is like a fast-moving river: a tumbling aggregation of selective information in specific areas.

YOUTUBE: BROADCAST YOURSELF

YouTube is another opportunity (or distraction, depending on how you look at it) to learn how to build a chicken tractor, compost inedible offal, or truss a bird for roasting. It's also a place to upload videos about your work, whether you taped a speaker who came to your community or want to post a 3-minute video about your poultry program. Here's where you direct customers or potential funders for your a nonprofit. And those YouTubes of Keith Richards's guitar solo in "Gimme Shelter," aren't too bad either.

PINTEREST

Best described as an online pinboard, Pinterest allows you to share favorite photographs, recipes, books, and the like, using images. Like Facebook, Pinterest is image-driven but more so. Posting photos of your farm, your chicken, your flock, your guest speakers, or happy faces eating your chicken generates interest, followers, and traffic. You can also follow organizations that are of interest to you, such as:

- Stone Barns Center — supports young farmers
- Sustainable Table — follow them for the Eat Well Guide

INSTAGRAM

This image-driven social media site makes any photo you take look cooler and better. Similar to Pinterest, you "follow" people or people follow you. There's no asking to be friends as there is on a personal Facebook page.

Staying in the Loop

There are many ways to keep your program, and the related issues, in the public eye. Here are a few examples.

Create a Google News Alert about your poultry program, mobile poultry slaughterhouse, humane slaughter — or your farm and your product, so you can know all the good things people are saying about you!

EXPAND YOUR REACH BUT STAY GROUNDED

"I might never have found my way to Polyface Farm if Joel Salatin hadn't refused to FedEx me one of his chickens."

— Michael Pollan, *The Omnivore's Dilemma*

When Michael Pollan, the food journalist, contacted Joel Salatin, a diversified farmer in Virginia, he asked Joel to FedEx some chicken to him. Joel declined, stating that in alignment with his principles he won't ship his food. Mr. Pollan was welcome to come to Polyface Farm, however, and buy it there himself. So began a historic relationship that would help drive and shape the local food revolution.

When you sell your chicken to a restaurant, list it online on a site called Real Time Farms. It's a crowd-sourced (information supplied by the public) national guide to farms, farmers' markets, and eateries. You sign up for free and list your farm or your restaurant. Add photos.

List what you do in free online local-food guides like the Eat Well Guide (www.eatwellguide.org).

Use social media to promote educational events, such as outside speakers or films, and generate an audience.

Share photographs of your activities on your Facebook page. If you have a local farm-to-school program, for example, go and talk to kids who may never get a chance to visit a farm. Bring the farm to the students. Or tweet what you're doing to show others what's going on and how they can get involved. To find a farm-based education center, check out the Farm-Based Education Association (see Resources).

The winter months are a perfect time to participate in and develop your food community, especially in your school system.

Local Grocer + Local Farmer = The True Cost of Food

On May 25, 2011, Jefferson walked into Cronig's Market to deliver a batch of his whole, freshly killed pastured chickens. Together he and Greg Pachio, the butcher, determined a wholesale price. Jefferson went on his way, and the grocery had its first batch of island grown, permitted chickens.

The plump, rosy pink chickens had everything over flaccid, gray-white factory-raised birds: taste, freshness, and price. Although Jefferson's birds were significantly more expensive, they'd lived a chicken-life and their death was taken seriously and with care. The animals contributed, nurturing the pasture while alive — pooping, pecking, and scratching — and afterward, with blood, feathers, and bones. The government had given its seal of approval to the safety and cleanliness of the process and the food. The farmer was paid directly, and these monetary transactions circulated within a short three-mile radius. And the owner of Cronig's, Steve Bernier, had a product that reflected his philosophy: Buy local and sell good food.

It took many pieces of this Jenga puzzle to stack up, each balanced precipitously on the next, so that those birds could reach the meat counter. The last and final piece, though, was still in question. Just because the local chicken made it to the grocer, would customers buy it? The goal of the poultry program was always for the farmer to access all markets: from farmstands to farmers' markets, restaurants to caterers, grocers to even schools. Local meat was not to be restricted to one or the other market, limiting a farmer's ability to make money.

What is the real price of non-agribusiness chicken and would it sell? Cronig's island-grown whole chicken cost $6.99/lb. Compare this to the least expensive whole chickens at $1.59/lb., organic at $2.89/lb., and "all natural air chilled" at $4.69/lb.

"To equate the chickens that I raise with the other chickens in the supermarket — it's just not the same thing," responds Jefferson. "My chickens spend more than half of their lives living on the grass, breathing fresh air. And you can taste that. I do everything that I can to minimize the stresses in their lives. The other birds that you can buy in the supermarket are based off of overall efficiencies. And those efficiencies are based off price point. My efficiency is based on an ethical system."

Local chicken has a marketing edge in these days of meat recalls. Traceable, transparent, independent, ethically raised, humane, environmentally responsible, safe, fair-wage food has a cost and it is real. But in the end, what a worthwhile, tasty, tasty chicken, it truly is. (See recipes, pages 110–125.)

WASTE NOT, WANT NOT

Birds raised on pasture add their own natural goodness to the soil. They are humanely slaughtered and processed on the farm to create good, safe food. The water and bird bits — inedibles such as feathers, blood, and offal — are collected and added to the compost. All these activities when managed properly can improve the soil's health, save money on waste disposal, and close the nutrient cycle. Nothing is wasted.

Public Health and Backyard Growers

Boards of health have a problem. It's the growing number of people keeping chickens either for eggs or for broilers. Backyard growers are also raising birds in areas that are not traditionally farmed or even zoned "agricultural." Zoning battles are underway across the country.

The challenge for boards of health is how to address safe slaughter in all these tiny "custom" flocks that are meant only for private consumption, in contrast to birds that enter the food supply by sale. Urban egg farmers may not have planned for when their hens aren't making cackleberries anymore. The best agricultural practices must be met whether you are raising three birds or thousands.

Chickens will attract pests and predators. Be a good farmer even when you live in town. Attend a Poultry Day (page 66). Go to your local agricultural department, read books, and talk to other growers and farmers.

Most important, plan ahead. When you order your broilers, also schedule their one very bad day. If you wait until the day before, expecting the Crew to show up and they can't, and your birds keep getting bigger and bigger, you're endangering the birds' health and well-being. The job can be done with a traffic cone and a pot of hot water, but it's not much fun.

6

The Path to a Permit

Not all Cease and Desist letters are alike; some are sent simply to scare the farmer while the agency has no intention of following with an enforcement action. Other times the agency fully intends to take action if the farmer continues to carry on the conduct warned about.

— Pete Kennedy, Esq.,
Farm-to-Consumer Legal Defense Fund

To BEGIN YOU MUST IDENTIFY the regulatory agency that licenses the slaughter and processing of poultry in your state. Most likely, one agency issues the permit but other agencies will have a stakeholding interest.

For example, in Massachusetts three agencies have an interest and a regulatory stake: the Department of Public Health (DPH), the Massachusetts Department of Agricultural Resources (MDAR), and the Department of Environmental Protection (DEP). The actual permit application and the license are handled by the DPH — it holds the keys to the car. MDAR cares about animal health, DEP about water. MDAR and DEP both share composting concerns.

Read and understand the regulations and any exemptions they include. This information should be accessible online. If not, you may have to request a copy of the regulations from the respective agency or go and get them in person.

Locate and print the permit application. Like the regulations, the permit application should be online. If not, request a copy of the permit application or go get one.

Apply for the permit, to the best of your ability. It is highly unlikely that your state has a specific permit for mobile slaughter units. If it does, you're one of the lucky ones! The permit will most likely be applicable to brick-and-mortar facilities and thousands (if not millions) of birds.

Suggestion: Include a letter describing your organization or business plan and the scope and goal of your poultry program, and explain that you are looking for license to slaughter and process poultry for commercial sale, not custom sale. (See pages 88–89 for an example.)

Units vs. Trailers: The Language of Mobile Slaughter

In the current argot the word "unit" is usually attached to mobile slaughter — as in a Mobile Slaughter Unit (MSU) for either four-leggeds or poultry. According to the Niche Meat Processor Assistance Network, at this writing there are Mobile Poultry Processing Units in California, Kentucky, Massachusetts, Montana, North Carolina, Vermont, and Washington. A unit, however, is significantly different from a mobile processing trailer (MPPT).

In a unit, the kill floor and quality control areas of a slaughterhouse are reproduced on the back of a large walk-on trailer. The equipment is fixed into place — bolted onto a floor and covered by a pop-up tent or roof. The processing personnel enter and work on the unit. Everything takes place on the trailer bed, from killing to dressing.

The MPPT is not a unit. The mobile poultry processing trailer is a complete set of equipment; each piece remains free of the others. It is mobile by virtue of its trailer, which requires only a small truck to tow it. The work of slaughtering and processing takes place literally on the farm — on the ground where each piece of equipment is rolled off the trailer, set up, used, cleaned, and then rolled back on again. Collectively it is referred to here as an MPPT.

Permit Catch-22s

If IGI applied for a permit, it meant that we accepted the state's view that mobile slaughtering was subject to the same regulations as a brick-and-mortar facility. But this failed to reflect two fundamental realities of small-scale ag:

- There are a relatively small number of birds, if any at all, being raised in a local agricultural community.
- In some cases, small-scale mobile processing is the only way to jump-start poultry production and must be affordable and managed with the utmost care, preferably by a trained crew.

The DPH's permit was for a building; our trailer moved. The sheer volume of animals the permit considered was way beyond our reach and probably would never reflect, even with considerable growth, the number of animals any one processor would slaughter within our geographical area. The sales figures for one grower on the permit used the quantity of 10 million as a demarcation.

Nevertheless, we needed the Department of Public Health to help guide us toward food safety. Not keep us out or tell us we couldn't. No one wanted anyone to get sick; besides, a bad chicken meal is bad marketing for a farmer. IGI's bar was always set higher than any slaughterhouse's (although in view of the number of abuses of animals, people, and food that come from slaughterhouses today, one could argue that that was no standard at all). But we always strove for highest quality of animal welfare, worker safety, and food safety, and we did it, quite literally, in complete transparency.

No Shortage of Advice

I shared this situation with some colleagues in the niche meat markets. One fellow told me, "Don't even apply, because once you do, you've bought into their regulations. Keep doing what you're doing until they stop you." That was uncomfortable advice to hear, because that strategy would have thrown the farmers under the bus when and if we were shut down.

Another colleague told me that we had to apply because it was the only way to go, "and if you don't you'll be held liable." Liable for what, I was never sure. Liable for committing chicken death? Another advised me to apply but not to apply as IGI; apply as me. But the problem was it wasn't me. I was the executive director of a nonprofit that never took ownership of the birds or the food. I didn't own the MPPT; IGI did.

There were also people who advised me: "Don't go for any permits — go underground." But underground is the worst way to run a program. Underground is illegal. Underground is no business model. We were a nonprofit working to change things and that meant confronting the system, working the exemptions, not dodging the regulations. Dealing in black-market meat opens a Pandora's box of things that shouldn't happen to animal, land, food, or man.

I eventually contacted the Farm-to-Consumer Legal Defense Fund, who thankfully always pick up their phone. Although they wanted to help, they weren't sure what to do either and directed me back to where I started — the state.

Working Within

No matter how odd or onerous the system was, the only strategy was to work within it, and that meant working the margins. The exemptions or the gray areas in regulations were opportunities. Local food ought not to become a 21st-century version of *The Jungle* (see page 33). If farmers and eaters promote subterfuge, there will never be any real dialogue, trust, or change in the system.

That said, however, there were days when the intimidation, pressure, and seemingly shifting rules of the playing fields among local and state agencies, regulators, nonprofits, farmers, funding, and public relations made me want to throw up my hands, reach for the box of chocolates and the remote control, and sit on the couch all day. But that really wasn't an option either.

The goal in developing any kind of sustainable agriculture that includes local or regional poultry/meat production is that the regulatory infrastructure should be inclusive, responsive, and reflect the needs of all farmers who want to raise animals and the consumers who want to eat them. Thus transparency, as uncomfortable as it is at times, is the only option. Being straightforward and not intimidated by the regulators is the only road to respect and trust on both sides of the food — the farmer and the regulator. And all that eventually translates into an authentic, reliable, safe, and tasty chicken dinner.

We Raise Permit #417's Flag

It took two or three tries to get through the permitting process. The second time around, Richard Andre from Cleveland Farm was on board. Richard used to wear suits with ties, and he's lived in places like London and Amsterdam. He moved to Martha's Vineyard with his wife and son to change his life, and change he did.

Richard understands spreadsheets and numbers. He appeared to have memorized the regulations and the exemptions verbatim. It was Richard who defended IGI's position to DPH face to face and fought hard for the farmers. Traveling to Boston, he sat in sterile conference rooms, drank tepid brown liquid passed off as coffee in disposable cups, and stood his ground. Always respectful, always persistent.

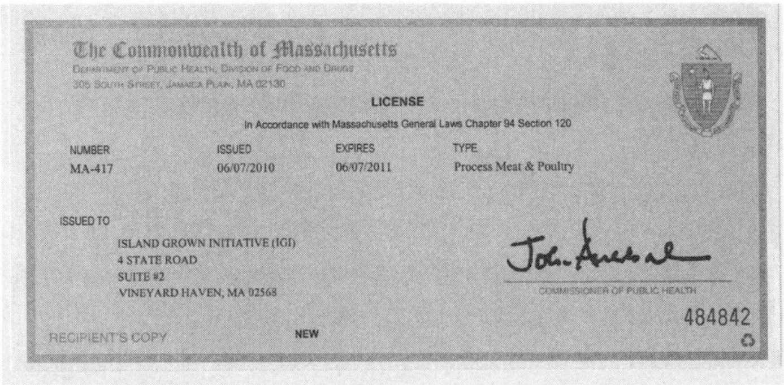
The Commonwealth of Massachusetts
Department of Public Health, Division of Food and Drugs
305 South Street, Jamaica Plain, MA 02130

LICENSE

In Accordance with Massachusetts General Laws Chapter 94 Section 120

NUMBER	ISSUED	EXPIRES	TYPE
MA-417	06/07/2010	06/07/2011	Process Meat & Poultry

ISSUED TO

ISLAND GROWN INITIATIVE (IGI)
4 STATE ROAD
SUITE #2
VINEYARD HAVEN, MA 02568

COMMISSIONER OF PUBLIC HEALTH

484842

RECIPIENT'S COPY

NEW

IGI's first permit

"For the DPH to get what they wanted," Richard said, "they had to have the farmers trust them. You can't beat trust into farmers. You have to gain it. So I asked the DPH, what outcome do you want? Because if you keep penalizing or intimidating farmers, you are going to get the opposite result of what you want. More farmers are going to sell underground chicken than are going to be upfront and transparent about it. It took the DPH a while to get that. But once they did, it was a watershed moment."

With Richard shepherding the MPPT as if he were a Border Collie–bulldog cross, IGI was finally granted Permit #417 by the Department of Public Health in the spring of 2010 under its pilot poultry program. IGI paid the $225 fee, and under #417's umbrella seven Vineyard farms in three different towns — The GOOD Farm, Cleveland Farm, North Tabor Farm, Morning Glory Farm, Flat Point, Northern Pines Farm, and the Whiting Farm —were allowed to process, market, and sell their chickens to the public via farmers' markets, restaurants, farmstands, and boardinghouses, but not yet in grocery stores.

Needless to say, it was a sweet victory. Even though we still hadn't obtained permission for farmers to sell at grocery stores, we were making progress. It was one step at time, and this proved a useful and successful strategy in negotiating with the state. The permit was the breakthrough that changed everything for farmers and for eaters.

In reciprocity, IGI negotiated that the farmers under the #417 umbrella had to register with the regional USDA office in Albany, NY, which included estimating how many birds they expected to process in one year. Under the USDA there are two cutoff points, one at 1,000 birds and another at 20,000 birds. At the time, all the farmers under IGI's permit #417 were below the 1,000-bird exemption level.

The farmers also agreed to sticker their birds with the USDA exemption number, IGI's permit number, the name of the farm, the location, the date, and safe handling instructions. Should the USDA want to drop in and check their accounting of bird totals, it could.

With Jefferson Munroe taking up leadership and management of the day-to-day MPPT logistics, IGI developed a Google calendar for the state and local boards of health, so they could visit any slaughter event any time they wanted to. Every farm that slaughtered commercially was visited at least once by a state inspector, and recommendations were made to each individual grower for improvements.

It wasn't always smooth sailing, but the pilot program worked. Glitches could be negotiated fairly because all parties were on the same page. That is one of the greatest beauties of transparency.

Exemptions

Generations have passed since Upton Sinclair exposed the stockyards of Chicago. Slaughterhouses have become mechanized havens of out-of-sight, out-of-mind efficiencies. Their assembly lines became speed models for those of Henry Ford. The consequences of mechanization demand regulations that reflect it. This isn't a necessarily a bad thing. Factory slaughterhouses need regulations. But regulations need to be adjustable down to the little guys. And that's essentially what regulators have done by including exemptions.

POULTRY EXEMPTIONS

"An exempt operation is exempt from all requirements of the Poultry Products Inspection Act (PPIA). Exempt operations are exempt from continuous bird-by-bird inspection and the presence of FSIS inspectors during the slaughter of poultry and processing of poultry products."

— Robert D. Ragland, DVM, MPH., Senior Staff Officer, Risk and Innovations Management Division, *Poultry Exemptions under the Federal Poultry Products Inspection Act.*

Those wise souls knew that all farms, and all kinds of slaughter and processing, are not the same. The world of agriculture and food is not one-size-fits-all. Although the Massachusetts DPH at first chose not to allow exemptions, Richard Andre had done his homework. He knew it was in their power to do so and he challenged them on that point.

"We propose," he wrote, "that a license that is scale-able and that does not disadvantage small farmers is appropriate and possible under current Massachusetts statute. Our proposal is to have the Commissioner of Public Health use the power vested to him under MGL Chapter 94 Section 130 to issue a license for exempt operations at a fee of $25. We would also support that the license is held by the individual grower/farmer. The current license fee of $225 is for slaughterhouses with

The Commonwealth of Massachusetts
Executive Office of Health and Human Services
Department of Public Health
Food Protection Program
305 South Street, Jamaica Plain, MA 02130-3597
(617) 983-6712 (617) 524-8062 - Fax

Application for Initial Licensure to Process Meat and Poultry in Accordance with M.G.L. C. 94, § 120 and/or 105 CMR 530.000

DIRECTIONS:

- Complete the entire two page application form.
- Submit a separate application for each facility to be licensed.
- Attach a separate check for each license application, made payable to: COMMONWEALTH OF MASSACHUSETTS.
  - $225.00 under $10 Million in sales
  - $375.00 over $10 Million in sales

1. Business Name:		2. Telephone #: ( )
3. D.B.A. (Doing Business As):		
4. Mailing Address:	Email Address:	
5. Facility Address (if different from #4 above):		6. Telephone #: ( )
7. Responsible Contact Person:	8. Twenty-four (24) Hour Emergency Telephone #:( )	9. Establishment # (if federally inspected):

Ownership	Name	Address
10. Individual		
11. Partnership	A.____ B.____	A.____ B.____
12. Corporation: A) President B) Treasurer C) Clerk	A.____ B.____ C.____	A.____ B.____ C.____
13. If Applicant is a Corporation:	A) State of Incorporation:	B) Date of Incorporation:

revenue up to $10 million. It would also be beneficial if the license period could be for a multi-year period.

"We also believe there should not be any additional restrictions to the amount of birds allowed to be processed under the USDA Poultry Products Exemption Act and its complementary state law. That being said, and there appears to be agreement across the various stakeholders, the current DEP limit of 2,500 birds a year is a reasonable level that requires the farmer/grower to enter into a discussion with both DEP and MDAR in order to formally review their water and composting plans."

IGI's strategy in implementing its poultry program was always grounded in the right to raise the food you want to eat to feed yourself and your family. "We (IGI) had brought Joel Salatin to speak to our community those few years back," Richard explains. "And it was powerful. He made an impact, at least on me. When you value the right to eat the food you want, it gives you courage to keep on fighting for that right." Thank you, Richard.

The DPH's Reason for Being

When I first read the application from the state's department of public health, it felt like a cruel joke on little local agriculture. The Application for Initial Licensure to Process Meat and Poultry (etc.) is the same one that Cargill, Oscar Mayer, and Perdue must fill out to open up shop in Massachusetts, if they choose to. The fee for the permit was $225 for "under $10 million in sales" or $375 for "over $10 million in sales." That was it: two sizes, big and bigger. Compared to what we were trying to build for our little farms, it was titanic.

We were to fill in these blanks, aware we were starting at zero:

- The plant will operate how many days a week? Hours per week? Hours per day?
- Estimate how many animals of each — cattle, calves, sheep, goats, swine, equine, chickens, capons, turkeys, geese, ducks, rabbit — would be slaughtered in a week.
- Estimate weekly volume of fresh meat or ready-to-cook poultry to be disposed in wholesale sales.
- Last, estimate the volume of product to be prepared and processed weekly — meats, sausages, edible fats, bacon, ham, fabricated steaks, poultry dinners, pies, canned meat or poultry, fresh cut or equine meat product, and other (specify).

I estimated one turkey a week and that was pushing it: I didn't know anyone at the time raising turkeys, but I didn't want to exclude it, should someone want to. I filled in the form, overestimating to the hefty number of 300 chickens a year or five or six broilers a week. I noted that we had no intention of humanely slaughtering other species, including horses, using the MPPT. But it was nice to know we could do rabbits if we wanted to.

Understanding Where They Come From

Making the DPH out to be the butt of the joke — or worse, the villain of this story — is not my intention: in fact, it is quite the opposite. The agency is charged with the regulation of safe, clean slaughter and processing in its mission to protect public health. It was simply operating under outdated and oversized regulations concerning the small to micro farm. The incredible thing, really, is that the agency acknowledged it and, in its way, engaged in dialogue and mediation.

Slaughter and processing make up one of those transformative systems that should connect farming and food.

The DPH also oversees the bottling of water, permits to immunize, burial and cremation, pharmacies, water systems, and food-borne illness. It is not an agricultural agency and it's not supposed to be. Along with the departments of agriculture and environmental protection they make up a system of checks and balances.

Nevertheless, there's no battle cry that resonates more deeply with farmers than this: "Keep your boards of health off my farm and out of my business." I've seen a room of farmers nod in brother/sisterhood and grumble a hallelujah chorus to this refrain. It seems everyone's got at least one nasty story to tell about how the board of health screwed over a farm in one way or another. I'd fallen for that siren's song as well.

But not anymore. Here's why. We're currently experiencing the sickening consequences of when a meat system essentially self-regulates and is monopolized by corporations or special-interest groups. Those groups' lobbyists in turn keep Congress on their side and in their pocket. So as difficult as developing working relationships across departments can be, it will make for more transparent and safer food systems in the long run, and that can only lead to safer meat and greater consumer confidence. Once regulations are responsive, scalable, and functional, they represent not just costs of doing business to the farmer but marketing opportunities to suspicious, salmonella-weary consumers, as well.

INTRASTATE OR INTERSTATE SALES

Intrastate sales: Sales within a state's border
Interstate sales: Sales beyond a state's border

In our tiny community, the demand for island-grown chickens far exceeds local supply. Jefferson Munroe of The GOOD Farm claims he can and does sell every chicken he raises. Along with other farmers he is permitted under IGI's license to sell only within the state. A permit that is limited only in interstate sales is not a problem for a farmer like Jefferson. As long as he can sell legally, that's the most important thing.

Whether intrastate sales are barriers to meat producers is being examined. A new cooperative program between the federal Food Safety and Inspection Service and state governments is examining the shipment of state-inspected meats evaluated at the state licensing level. As of 2012, Ohio and Wisconsin were on the cutting edge, with North Dakota coming up strong. This new initiative may prove to be one of those instances where regulations can change to support more local and regional food systems.

Cease and Desist

The day had started early, as usual, and events were going along pretty well. But I'd been rattled since the day before when, at a farm-to-school conference, I was accosted by a USDA agent. For privacy's sake, and in the spirit of the acronyms that our government so loves, I will refer to him as Rude Man in Gray Suit.

"Are you Ali Berlow?" RMGS asked tersely as he presented himself with nary an outstretched hand. I remembered we'd met before, at a conference about local and regional meat production. But before I could chit-chat about farm-to-school initiatives and school lunch he blurted out, "So what about the Cease and Desist?"

"Excuse me?" I replied. Knowing nothing about any Cease and Desist order, I went into predator-alert mode.

RMGS was certain that I or IGI had been served a Cease and Desist order regarding the mobile trailer. "Everyone's talking about it," he said. "What are you going to do now?"

My heart and stomach both wanted to jump out of my body. My mind raced, wondering if I could have missed such an action. But don't you have to sign a receipt for something so draconian-sounding? Nothing like that had come through our mail — I was sure of it.

"Gee," I said. "If everyone's talking about it, then I wish someone would talk to me about it. Sorry, no Cease and Desist here. May I have your card?" Mr. RMGS turned out to be USDA Rural Development Director of Community & Business Programs.

I quickly found myself a quiet room where I could make some phone calls. First my husband: "Honey, we haven't gotten any special mail recently, have we?" No, nothing had arrived by certified mail. Then I called Jim Athearn of Morning Glory Farm, where we'd scheduled Flavio and the Chicken Crew to attend to Jim's flock the next day. I feared the said C&D had been delivered to him and that the RMGS simply had his facts wrong. Jim, steady and even-tempered as always, appreciated the check-in. All was well and on track for the slaughter.

WHAT'S A CEASE AND DESIST?

Black's Law Dictionary defines Cease and Desist as "an order of an administrative agency or court prohibiting a person or business firm from continuing a particular course of conduct."

A C&D could be issued by an agency or a district attorney. It is a quit conduct of behavior, as issued by an administrative or judicial action. In the real world and practically speaking, there are two kinds of cease and desist: One is essentially a warning shot sent across your bow with no real intent to follow up with enforcement. The other is more serious and is issued with the intent to follow up.

You will know when you receive a C&D. It will be delivered to you by certified mail or handed to you in person.

You are allowed to request a hearing to learn about its origin. It's done on a case-by-case basis and under the interpretation of what the agency or inspector intended.

It's best to contact the Farm-to-Consumer Legal Defense Fund if you've enjoyed the pleasures of a C&D. They have a national perspective and a sense of the regulatory temperature in the country and, hence, in your state.

My last call was to Flavio, and it was my most dreaded. Flavio is Brazilian: a legal immigrant but an immigrant nonetheless with a strong accent and honey-colored skin. I feared for him. I feared that he was easy pickings, the most vulnerable link in this chain, and that I had made him vulnerable to intimidation by our government. I felt responsible.

But all was well with Flavio and his wife Marcia. At that very moment they were in my driveway in Vineyard Haven, where the mobile slaughterhouse was stored, hooking it up to drive it to Morning Glory Farm. It's our practice to get the MPPT to the location the day before. One less thing to attend to on the Big Day Of.

No Cease and Desist had been served that day. Rattled and disoriented from the encounter, however, I left the farm-to-school conference early. I could only think that bad things were underway that might throw the MPPT, the farmers, and the poultry program under the bus.

Once home, I called my lawyer to describe the face-to-face encounter, and that's when I finally got emotional and upset. An agent of the USDA is an agent of the federal government. Why use intimidation when you're already wearing the badge? My lawyer was calm and gave me what may be one of the best pieces of advice I've ever received about being an activist.

I could only think that bad things were underway that might throw the MPPT, the farmers, and the poultry program under the bus.

"Welcome the Cease and Desist," Frank said. "Welcome it?" I answered, confused. "Yes," he said. "It's how the system engages." Whether it is the Department of Public Health or the USDA that is handing out the Cease and Desist, it's just another move on the chessboard.

This helped solidify my understanding that these agencies by definition can only be reactive. You have to the throw the ball against the wall to see, react, and catch it after it bounces back at you. That's how you play catch with regulations. Otherwise, nothing will ever happen. Nothing will ever change. It's another reason it takes a strong constitution to advocate, to stand there and take whatever the RMGSs of the world, God bless them, are trying to dish out.

7

The Big Day Of

I loved the truth. Even in just this one thing:
looking straight at the terrible,
one-sided accord we make with the living of this world.
At the end, we scoured the tables, hosed the dried blood,
the stain blossoming through the water.

— excerpt from "What Did I Love"
by Ellen Bass. Used by permission.

PLANNING FOR THE SLAUGHTER starts the day a farmer's chicks arrive in the mail, because that's the same day the Chicken Crew should be contacted to schedule the MPPT. In those 8 to 10 weeks as the chicks grow out, try to connect a farmer new to the MPPT with a farmer who will be slaughtering in that time period. Seeing the MPPT and the Chicken Crew in action will help new farmers understand how to prepare their property for their Day Of.

See page 127 for the Farmer's Checklist. The farmer should have received this list for preparing her farm well in advance of her Day Of. Let there be no (or at least few) surprises on the chickens' one bad day.

The Farmer's To-Do List

The day you start rearing your broiler chicks, schedule the slaughter dates with the Crew as mentioned above. As the birds grow you can make a calm, smooth, transparent preparation for their final day.

A Few Weeks in Advance

Choose the site(s). Provide access to power and water. Assess prevailing wind in relation to food crops and produce storage areas. Avoid road dust and overhanging trees. Have your drinking water tested for total coliform and nitrate/nitrite; have a copy for the state and a copy for your local board of health.

Obtain two copies of your *Salmonella pullorum* certificate: one for the state, one for the board of health. This is a very contagious bird-to-bird, hen-to-chicks disease. For more information, go to the USDA's Animal and Plant Health Inspection Service site and specifically The National Poultry Improvement Plan (NPIP).

Obtain plenty of wood chips or other absorbent material to absorb runoff. Do not use sawdust or shavings as they are too light and can get swept up by a breeze.

Set up your composting in accordance with requirements of your departments of agricultural resources and environmental protection. To find regulations in your state, search for "on-farm" + "composting" + (your state) + "guidelines" or "regulations." The Cornell Waste Management Institute of Cornell University is also a terrific resource on composting livestock and "butcher waste."

A Few Days Before

- ☐ **Mow and clear the site** of natural and manmade debris.
- ☐ **Test your fridge and freezer,** if you expect to hold any birds at the farm, and calibrate your thermometer.
- ☐ **Clean and make space in freezer/fridge.**
- ☐ **Clean and sanitize coolers** you will use to pack the finished, packaged birds.

SANITIZING

Use bleach water to sanitize equipment: one tablespoon to a gallon of water. Use test strips to test the bleach concentrate. Use only straight-up bleach, no additives, no scents. You should not be able to smell bleach.

☐ **Clean your vehicle** for transport: rinse all visible contamination, scrub if necessary, and sanitize prior to transport.

The Night Before

☐ **Take the birds off feed** but allow them access to water. This will help clear their digestive tracts. (See box, page 98.)

Morning Prep

☐ **Remove or restrain** free-range livestock and pets.
☐ **Place wood chips** below the kill cone, plucker, and evisceration table. Note: if there is a grade from kill-side to evisceration side, LOTS of chips will be required.
☐ **Get birds to the site promptly,** without injury, and provide a shaded area for them.

During Slaughter

☐ **Assist the crew** with bagging, weighing, and labeling the finished birds (see pages 59–60).

LABELING

If you are processing fewer than 20,000 birds a year, your farm sticker should include the USDA Poultry Exemption citation: *Exempted page PL.90-492.* You can make your own labels on your computer.

☐ **Properly pack bagged birds in ice** and continue to monitor their temperature.

After Slaughter

☐ **Oversee final disposition** to consumer, refrigerated holding, or freezer.
☐ **Oversee site cleanup** and proper composting/disposal of remains.
☐ **Monitor holding temperatures** of stored chickens per protocol.
☐ **Transport chickens** to the buyer in sanitized closed totes, packed in ice, properly secured.

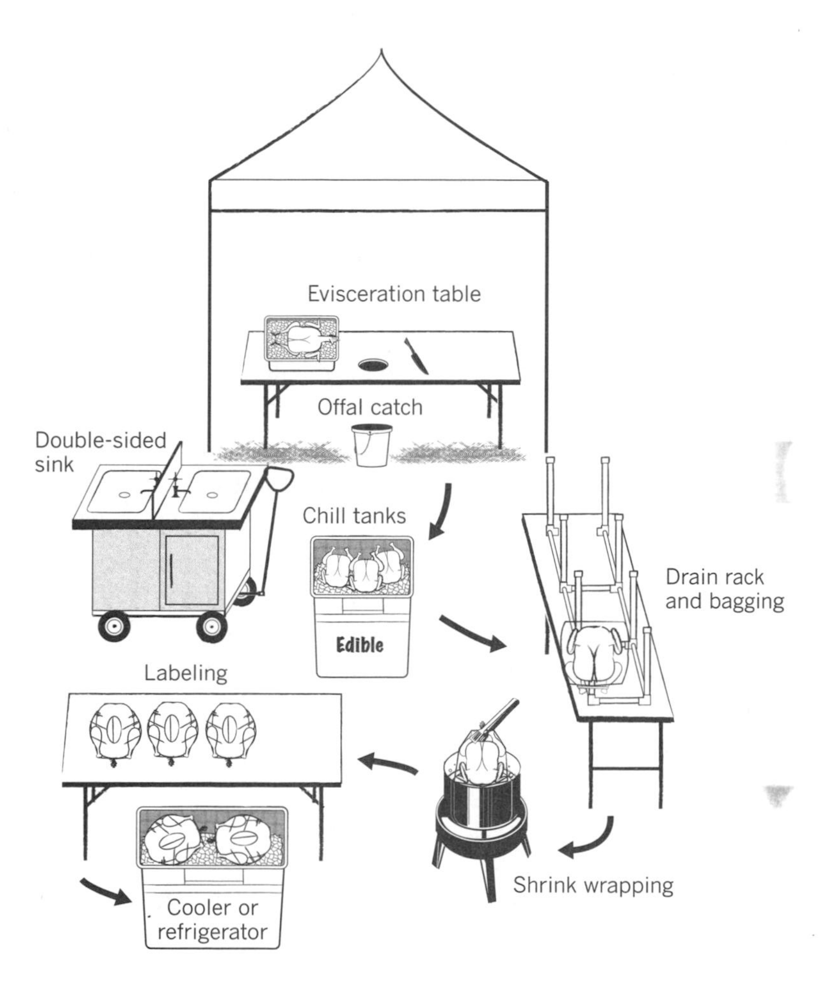

Close-up of the Clean Side: evisceration and then bagging. Each of these processes is best kept under a separate tent.

Advice to a Farmer from the Chicken Crew

Jefferson Munroe offers the following tips:

As the farmer, if you can assume an observer's perspective (as opposed to a doer's perspective) during the day, this is extremely helpful for food-safety oversight.

All farmers should take a ServSafe course (see page 53) or at least a low-level slaughter-specific food safety class for farmers. Ask your regulators if they can offer this to you. Farmer food safety education supports and builds confidence in regulators, Crew, and customers. It also builds communication and understanding between the Crew and the farmer.

Plan ahead for reliable refrigeration. A typical home fridge (33 inches wide by 66–69 inches high by 32–25 inches deep) can hold about 35 birds at 34°F (1.11°C).

Ideally, the MPPT will have access to hot water. This will help heat the scalder faster so it will use less propane. Hot water is also helpful if you need to add water to the scalder throughout the day, and it makes cleaning the MPPT much easier.

Mow the grass where the Crew will be working. It's lovely to imagine working in a hay field, but it's good to be able to see the ground.

Spread wood chips under the kill cones, plucker, and the evisceration table. Do NOT spread wood chips under the scalder.

Prepare lots of ice, more than you could ever imagine. Not big blocks. Smaller is better. Shaved is fine if you go by weight and not volume.

Stop feeding the birds ahead of time. The recommended time is 8 hours, but it's better if it's closer to 12 to 15 hours, especially if they are on pasture. For example, if you're aiming for an 8 AM start, then pull feed around 3 PM the day before, as birds' systems quiet down during the night and will take longer to digest.

WHY PULL FEED

If they're on pasture — as in living in mobile chicken tractors with access to grass all the time — then to "pull feed" means to move the birds to fresh grass around 3 pm the day before. Do not give any grain or dry feed, but do have all the water they can drink available to them. Water aids in digestion and helps keep stress levels down.

There is a big difference in processing birds that are on pasture with some grain and birds on an all-dry-feed diet. Because grass is relatively wet, if there is some sort of breakage in the intestinal tract during gutting, the grass is less likely to get embedded in the carcass. In instances of breakage, wash, rinse, and sanitize processing area and utensils for one minute.

Minimize crate time. Collect and crate the birds as close to slaughter time as possible.

Gently catch and crate the birds one by one. Grab–collect–scoop up a chicken over both sides of its mid-section and pin in its wings. It's useful to have a second person to help open and close the crate as the birds are being caught.

Never catch a chicken by one leg. If you must go for legs, always go for both.

Keep the birds shaded when they are in their crates. Being in the crates is stressful enough. Keep them cool.

Wash your hands. In particular, always wash your hands with hot water and soap when you move across the threshold between the kill side and the evisceration side.

Allow the meat to rest in refrigeration for at least one full day before freezing. It makes for a better-quality chicken.

Schedule delivery and distribution of your birds to markets (restaurants, grocer) the same day you process because: 1) there's an inherent risk that your refrigerator may fail; and 2) the birds take up a lot of space.

Don't sell directly to home cooks the Day Of. Or if you do, have them wait 12 hours to one day before cooking their fresh and humanely slaughtered bird. It will still be going through stages of rigor mortis, and this natural process is slowed when the bird is chilled. If they eat it the Day Of, they'll experience a tough bird, which will reflect badly on you and your product. Farmers, note: according to the Massachusetts DPH, a freshly killed refrigerated chicken should be sold or frozen within 4 days. A frozen vacuum-sealed bird will keep for up to 6 months in a −20°F (−28.9°C) freezer.

CHICKEN CREW DOCUMENTATION

- Oversee the time/temperature testing and documentation of carcasses and giblets using Clean Technique: sanitize the probe between birds, and sanitize hands and pen before and after the documentation.
- Document the number of birds falling out of the kill-cone and other events on the kill side.
- Document ingesta-spill, gall-bladder rupture, equipment contamination, and mitigation during evisceration.
- Do a final inspection of birds out of the chill tank prior to packaging, and document any problematic findings (broken wings, sick birds, contamination, and so on).

Advice to a Farmer from a Local Board of Health Agent

by Marina Lent, Chilmark, Massachusetts

Marina Lent, Chilmark Board of Health Administrator, worked closely with her colleagues in the six town boards of health in our community and with the state office in Boston. With a commitment to a strong, safe, and local food system, Marina embraced the MPPT with vigor, determination, knowledge, and all the caring one could hope for. That she works for the town government is more than fortunate. Marina graciously offered three tips to all parties for public health and food safety, from her position as a board of health administrator. It demonstrates an inspector's perspective and priorities, in the language and realm of public health. Here are Marina's "Three Principles of Safe Slaughter":

1. Prevent contamination of the edible bird.
2. Maintain time/temperature controls to prevent bacterial growth.
3. Prevent contamination of your farm.

Gray Water Awareness

The departments of public health and environmental protection expressed great concern about the wastewater from the slaughter. The state was asking for elaborate water collection devices or traps so nothing would hit the ground — ice melt, water from the scalder or plucker, hoses used in gutting and dressing the birds.

LESSONS LEARNED ABOUT WATER

- Farmers, have your water sources tested for potability and have copies of certification on hand.
- Select on-farm location(s) for the mobile slaughterhouse that are away from water sources like ponds, lagoons, streams, and wells.
- Chill tanks are cold. Investing in some leg hooks to pull chickens out of tanks will make the Crew happy.
- Be prepared to shovel, scrape, rake, or hoe the post-processing water-soaked woodchips and dump the used water from the scalder and the chill tanks into your compost.

THE ELEGANT SOLUTION: WOOD CHIPS

Wastewater from the slaughter day is of concern to regulators. The wastewater from the plucker, scalder, and chill tanks will contain biological bits of bird: there's no preventing that. With proper management this water can be collected and added to the active on-farm compost that should be a part of your poultry program. (See Lessons Learned about Water, box on page 100.)

Spread clean wood chips (never pressurized wood chips) under water-intense areas in the MPPT setup: kill cones and especially the plucker and the evisceration table, but never the scalder (that would be a fire hazard). The wood chips soak up the water and create a safe work surface for the Chicken Crew. After the day's events, when the equipment has been washed down, shovel up the wood chips and put them into the on-farm active compost. (For composting information, see Resources.)

In the early days of running the MPPT, at a slaughter on Morning Glory Farm, I'd asked a local water expert, Elmer Silva, to observe what was going on and suggest any good ideas about water collection, this sticking point with the public health department. Elmer is an engineer, had worked for the town waterworks, and runs a successful pool and hot tub business on the island. He knew how to collect and move water and wasn't squeamish about slaughter either.

He looked around the farm while Flavio and Crew carried on their business. I laid out for him the health department's concern with all the water. "Should we have the Crew stand on top of some kind of kiddy-pool thing?" I asked. "Or modify a Slip 'N Slide with a safer surface to stand on and a gutter to collect the waste water into a container?"

He looked at me as if I was crazy. I felt crazy saying it, too — endangering the crew that way. I was trying to collect this gray water in a way that would satisfy the regulators and still make our operation functional and realistic. The water was dirty, but it was dirty with natural, biological bits of animal. You don't call in the Department of Environmental Protection to permit road kill. How bad could 100 chickens be when you're composting the bulk of it? And how much water does your neighbor use washing his car in the driveway every weekend — that's runoff that goes into our sewers and lakes, rivers, and oceans, and no one's making him get a permit for that ... but regulators don't really like to hear those arguments.

"A couple of things," Elmer said, after watching for a while. "Don't set up near a stream or a pond. And you don't want the Crew standing in water, so set up on slightly higher ground so the water runs off and doesn't pool around them. But about collecting it? This is a farm. Things happen on a farm. They may not be pretty and they don't happen in a city. But it's a farm."

Enough said. We took his advice and kept on going, kept on learning.

How Many Chickens in a Day?

by Jefferson Munroe

A long day is 125 birds, once everyone's trained up. Try not to do more than 100. Have three trained crew members plus a helper (the farmer or a volunteer). If you're set up and ready to go by 8 or 8:30 you should be done with 100 birds by noon.

Calculate that one Crew member can do about 25 birds in a day, from setting up to breaking down. So 75 birds should have three people, 100 birds four people, and so on, since there are attendant tasks outside the killing and eviscerating. If there's a bottleneck in the work flow, it's going to happen at the evisceration table.

Having three Crew members is ideal from a labor standpoint: one on the kill side, two on the processing side. With less skilled folks, 25 birds per person per day is a good rule of thumb to avoid situations where time becomes an issue. No one wants hand cramps from eviscerating chickens for seven hours.

On the kill side, a skilled Crew will take 1 to 1.5 minutes per bird from killing to scalding to plucking. (The scalder fits three or four birds at a time). That's 100 to 150 minutes for 100 birds. On the evisceration side, a skilled Crew can do 20 birds an hour, or a bird in 3 minutes.

We've found that with a proper ice slurry and pre-chill tanks we can lower the temperature of birds under 6 pounds (dressed) to below 41°F (5°C) within 2.5 hours. Bagging can be done at our own leisure at that point, as long as chill tanks remain below 40°F (4.4°C) throughout. With a trained crew of three or four people we can process 100 birds in about three hours from kill to chill tank. (There's a bit of lag time on both ends of the process: when the kill side gets going and while the residual birds are eviscerated after the kill side is finished.)

When the evisceration side is really kicking we can do around 40 birds per hour for a few hours, but we generally average closer to 30 to 33 per hour when temp taking and monitoring is factored in. I can eviscerate a bird every 2.5 minutes once the season gets going.

If you're doing more than 100 birds, it's helpful to have a fourth Crew member to get birds through faster.

TALKING TURKEY

Turkeys take much longer, about 40 minutes per bird from kill cone to chill tank, whereas a chicken takes 8 to 10 minutes. The chicken minutes include killing, plucking, and eviscerating time that generally requires two people — one on the kill side and one on the eviscerating side — plus bagging time. Turkeys, being larger birds, usually require three to four people on the kill side, and each turkey has to be individually killed, scalded, and plucked in this type of MPPT. This vastly increases the person-hours required for the Thanksgiving dinner.

If it's a commercial processing, extra time is $4.25/bird to monitor and keep records to ensure that all food safety precautions are being taken.

For example, here are the costs of processing 100 birds with the Crew manager plus two workers (three Chicken Crew total):

- $135 MPPT rental fee
- $90 for ice. In this scenario, ice = 20 cents / pound. Each bird requires, on average, 4.5 pounds of ice but more ice is always better). That equals: 90 cents of ice per bird multiplied by 100 = $90
- $425 = labor

Cost to process 100 birds under license = $425 labor + $135 rental + $90 ice = $650. Cost to process 1 bird under license = $6.50

When Things Happen: Advice from an Advocate

There is no doubt about it, Day Ofs are stressful. You'll be sure you've forgotten something critical or missed something obvious. I never knew if we were going to be greeted by an inspector, an officer of the court, an animal control officer, an animal welfare activist, an upset neighbor, or a dog off-leash. Best thing you can do is to breathe.

Many things can and will go wrong on the Day Of. Arguably, one strength of IGI's strategy in developing its poultry program is having a trained crew dedicated to the jobs at hand. If there's not enough ice, the propane tank runs out, or the scalder's auto-timer is broken, it's up to the Crew to troubleshoot potential problems and address them professionally. And if the farmer's cows get out or she goes into labor, she can attend to that and not stress about the slaughter. With a trained crew, everyone involved can focus on their priorities and responsibilities.

As organizer and advocate of the program, especially in its early stages of its development, your job is to ensure that the Crew, the farmer, and the animals are all safe and that the Day Of results in high-quality food produced, waste managed, and people paid. It's best to be able to anticipate. Be ready to grab any arrows shot your way right out of the air before they hurt the farmer, your poultry program, the farming community, and any other mobile humane slaughter programs out there. Put on your figurative armor or invoke your white light, however you want to think about it. At the end of the day you can wash off it all off in a hot shower with some good-smelling soap.

The point is to be on point. Anticipate.

Unexpected Visitors

If anyone comes to the farm looking to stir things up — an animal-rights activist, curious journalist, health inspector — be polite, introduce yourself, shake hands, and ask for his or her card.

BE PRESENT ON SLAUGHTER DAY

Be open, alert, calm, mindful, and ready on slaughter days. They are truly amazing. You're working outside, on a farm, and supporting the farmers who raised their birds and are now taking responsibility for the humane slaughter of their animals as well. This is a gift that few people get to experience. These days are honorific, a testament and proof to life cycles, and you get to be part of it. The culmination is pretty damn awe-inspiring — safe, quietly respectful, and full of dignity. Your job as an advocate is simply to keep it that way.

Put yourself between him or her and the Chicken Crew. Once the day starts, the Crew needs to keep focused and working. Be respectful and firm in your convictions and your knowledge.

Ask your guest what he or she is there for. Is she with the press, is he a concerned citizen or neighbor, an inspector, a friend, or a foe?

If they're there on official business — such as to shut you down, as in a cease and desist (see What Is a Cease and Desist? on page 92) — ask them to hold on a minute. Let them know you're calling your legal defense and then the newspapers. You can even ask if they'd like to give the reporter a quote.

If and when the Crew's workflow allows, ask that the Crew manager be available to meet and speak with your guest, if that will help.

Not every Day Of will present itself with these types of challenges. But it could.

Other Tips for the Advocate

Charge your phone.

Pre-program your speed dial with:

- The Farm-to-Consumer Legal Defense Fund hotline: 1-800-867-5891
- One or two local reporters — television, radio, or print
- One or two regional reporters

Bring a camera, charged and ready to go. If it has audio and video, all the better.

Do not offer regulators or inspectors anything to eat or drink. It could be misconstrued as a bribe.

Enlist a friend or neighbor to document the activities.

Have a Facebook Fan Page and/or Twitter account (see pages 75–79) and don't be afraid to use it.

If you are presented with a Cease and Desist, alert and inform the Farm-to-Consumer Legal Defense Fund. They will help you follow up.

If you're facing trouble and need your supporters to rally around you quickly, sound the alarm that you need them to show up and make their voices heard. Don't cry wolf, but do cry for help.

Know your rights and don't give them up.

WHY WEAR PINK

I wear pink on slaughter days. I wear pink so I can be found. I want farmers, regulators, animal rights activists, and reporters to talk to me. I want to be found so I can do my best to catch any arrows in midair before they hurt farmers or blow a hole in the program.

Plus, pink is a happy color. It makes people smile. Even when there's not-very-nice work to be done.

Custom Chicken

"Burn it!" the customer said loudly over the yellow flame dancing from the blowtorch, her voice bouncing off the echoey walls of the slaughterhouse. Chickens could be heard squawking in the room next door.

"Like this?" Maria shouted back, pointing to the pecan-colored skin of her forearm, out from under her coverall.

"NO! Like this!" the customer shouted back in a smile. White teeth parted by red lips on shiny eggplant-purple-kissed skin. Pointing to her face: "The color like *me*!"

Maria torched the dead naked bird with that cool flame until all the feathers were burned off, and then a little longer and a little closer to the bird, until the chicken skin turned black, stretched taut, nearly splitting and torn but not quite.

The customer, who is always right, was from Cameroon via Connecticut, and she never buys meat from the grocery store. "This is the way we have it back home," she explained. "Black. That taste of fire. That is what I want. And it is so good!" She was rapturous to have this taste of her faraway home right here in America.

Juliet found her way to this slaughterhouse along with many other immigrants to the New England area. It is a small, permitted slaughterhouse that does custom slaughter only. The meat that comes out of these doors is never to enter commerce. It cannot be sold or bought from a market. You won't find this slaughterhouse on the Web or in a phone book. It's on a farm. They don't need to advertise; their customers find them. You would not know it's there if you drove by. It's a rare, nearly extinct species — a throwback to the days when small towns and neighborhoods across the country had slaughterhouses.

This custom house also caters to the tastes, traditions, and religions of immigrants who come from all over the world. By doing so, it sheds light on the culture of the slaughterhouse, its laws and regulations. Whose value system do we adopt when regulating a slaughterhouse, and what does meat mean to us? Are we conventional or sustainable agriculture, local, or somewhere in between? Are we religious, atheistic, humane, mechanical, or other?

All Juliet wanted was lean fresh chicken that tasted like home.

The very next day marked the beginning of Eid al-Adha, a most holy Muslim holiday, and hundreds of families — American, Asian, African — would arrive at this place for the ritual sacrifice of lamb and the halal meat that would grace their holiday table. The imperative is so strong that these ritual slaughters will go forward even if places like this get written off the regulatory map and no longer exist — moving to backyards and back streets.

The birds Juliet bought were spent laying hens from a large egg-producing factory farm farther south. They most certainly had never pecked and foraged on a green grass pasture. This, however, was of no consequence to Juliet. She seemed not to have an iota of concern about where the chickens came from, what breed they were, or what kind of lives they led. The most important thing to her was to see the bird alive one minute and dead the next. All she wanted was lean fresh chicken that tasted like home.

Juliet bought her usual share, six chickens to get her through the next few weeks. She left with her gutted and loosely bagged birds, flirting and waving jovially to her slaughterhouse friends. She'd see them soon enough when she made the day trip for her black-skin chicken again.

8

Now Cook Up That Lovely Bird

Winner winner chicken dinner!

— Traditional exclamation when someone wins a prize

The best part of the MPPT is how community comes out of the woodwork. Everyone loves chickens in one way or the other. The fact that you're out there, making something happen for eaters and farmers, is such a positive thing. Have fun and enjoy the people you meet. In so many ways, that is the gift of this work. Like a good tale that has a beginning, a middle, and a happy ending, this journey's stories all start in the kitchen, and back to the kitchen they return.

Making the Most of It

A local chicken is different from typical grocery chicken in size, shape, proportions, and texture. To get the most out of your bird, cast a wide net into different cooking techniques: roasting, braising, stock making, salads, soups, stew. Learn how to carve well and how to get in there to pick all the meat off the bones. Hone your skills in how to butterfly a whole bird for the grill or to put up in your freezer. Save all bones and skin for stocks.

The following recipes, contributed by Jefferson Munroe, Robert Booz, Daniele Dominick, Betsy Carnie, and Gordon Hamersley, were selected because these folks all know their chicken, and because a local chicken — one raised in small flocks on the pasture, an ethical chicken, as it were — is going to be more expensive. So in the kitchen it's crucial to make the most of the beautiful bird you have before you. It is healing to cook with a good, fair, fresh, and safely and humanely killed chicken. And best of all, it tastes true!

TYPES OF CHICKEN

- **Broiler and Fryer.** 7–10 weeks old, 1½–3½ pounds
- **Roaster.** 10–14 weeks, 4–6 pounds
- **Capon** (castrated rooster). 5–6 months, 6–12 pounds
- **Stew Hen.** 10 months–1½ years old, 4½–7 pounds
- **Rooster/Cock.** 10+ months, 4–7 pounds

Note: Ages and dressed weights depend on breed, the life lived, the feed fed.

Labels: A Matter of Words

contributed by Robert Booz, sustainable food activist, writer, cook, hunter

Food labels can be bewildering, the terminology open-ended and even poetic, yet crafted to obfuscate. Know what's marketing and what's real.

Basted or Self-Basted. As defined by the USDA Food Safety and Inspection Service, poultry to which "fat broth stock or water plus spices, flavor enhancers, and other approved substances" have been added must be labeled as basted or self-basted. It is not uncommon for poultry processors to baste birds to add weight to the packaged product.

Free Range. The USDA Food Safety and Inspection Service allows poultry to be labeled as Free Range or Free Roaming if producers "demonstrate to the agency that the poultry has been allowed access to the outside." What is problematic about this terminology is that the amount of access or the relative type of access is not regulated. Overcrowding is still prevalent in many cases, and the birds do not necessarily ever make it outside.

Local or Locally Grown. These terms have no legal definition. Eating locally is important, but there is no regulation around the use of this term. One hundred miles is a popularly accepted range of local foods, and 300 miles is popularly accepted as regional; however, this varies from location to location. What's more, just being local is no way to ensure the quality or sustainability of an ingredient, so consumers must investigate for themselves.

Natural, All Natural, 100 Percent Natural. These terms have no legal definition and no basis in growing or production methods. Almost anything can carry this label.

Organic. To be labeled organic, products must meet the standards set by the USDA National Organic Program. Organic farmers must use organic (non-GMO) feed and must not use antibiotics or growth hormones. This does not necessarily mean that the bird has been raised in a free-range environment.

Starting with the Whole Bird

Roast Chicken Permit #417

Roast chicken was the first meat my older son ate. My neighbor Maggie had cooked up a beautiful bird. The interaction between the cook and the baby — their pleasure — is one of my most enduring kitchen memories. Now both my sons, all grown up, work part-time on the Chicken Crew in the summer, giving this recipe even deeper meaning to me. It serves 4–6, with leftovers.

- Brine (see page 112)
- 1 whole local chicken, 3–6 pounds
- 1–2 carrots, peeled, cut in 1-inch-long sections
- 1 onion, roughly chopped
- 4–12 peeled garlic cloves, smashed
- 2 celery stalks, cut into 2-inch lengths
- 1 lemon, quartered
- Thyme, fresh, a sprig or dried, about a tablespoon
- Duck fat, chicken fat, butter, or oil
- Kosher salt, fresh ground pepper

Brine the chicken for 4 to 24 hours (for basic brine instructions, see page 112).

Preheat oven to 450°F. Rinse the chicken and dry well. In a size-appropriate iron skillet or roasting pan arrange a bed of the carrots, onion, garlic, celery, and a couple of quarters of the lemon. Place a piece of celery, some onions, a quarter or two of the lemon and the thyme loosely into the cavity of the bird. Mix the fat of your choice with a good bit of salt and pepper and rub it under the skin of the bird. Rub a little more on the surface of the skin.

It's not strictly necessary that you truss it, but it's a time-honored tradition in the arts of cookery that tightly binding the legs and wings to the body ensures moister breasts and more even cooking. Additionally, a trussed chicken will appeal to cooks and diners who share a particular aesthetic; elbows in, legs crossed, neat and lady-like. So as you eye your raw chicken, wondering whether or not to bind it into a tight ball of bird, I offer you Molly Stevens's smart and simple suggestion found in her inspiring cookbook *All About Roasting: A New Approach to a Classic Art* where she writes: "Ignore any complicated cat's-cradle-style trussing instructions and simply tie the two drumsticks together with kitchen string. This approach gives you a prettier bird that roasts as evenly as an untrussed one."

Place the bird on the bed of vegetables. Roast the bird uncovered in the hot oven until you can smell it and it's turning brown, about 20 to 25 minutes. Turn the oven down to 350°F. Continue roasting for 45 to 60 minutes. Cooking time depends on weight of bird (see box below) — usually 20 minutes per pound.

Test for doneness by jiggling a drumstick. It should be loose in the joint. Juices should run clear if skin is pierced with the point of a sharp knife. If you use a meat thermometer, insert it in a leg without touching the bone. It should read 170°F.

Remove from oven, cover loosely with foil, and let rest for 10 or 15 minutes. This allows the juices to return to the meat. Carve and serve with some of the roasted vegetables.

Save the carcass, any fat, juice, and vegetables (except the lemons) from the pan and any leftover bones on plates for a quick stock.

COOKING TIMES AND WEIGHT

Here are some approximate cooking times for roasting pastured, unstuffed chickens:

- 2½–3 pounds: 1¾ hours
- 3½–4 pounds: 1½–1¾ hours
- 4½–5 pounds: 1½–2 hours
- 5–6 pounds: 1¾–2½ hours

The Brine

by Robert Booz

Brining is one of the most important and yet overlooked things you can do to a chicken. Everything from fried chicken to grilled chicken to roast chicken is improved with a little soak in some brine. Here's a simple brine recipe anyone can make:

- 1 gallon cold water
- ½ cup kosher salt
- ½ cup light brown sugar

Combine the ingredients. Bring the brine to a boil, let it cool, pour it over your chicken, and let it sit in the fridge, preferably overnight. If you are short on brining time, make the brine saltier; if you are going to let it brine for a few days, cut back on the salt and sugar a bit.

Trying throwing in aromatics like garlic, allspice, peppercorns, juniper, herbs, maybe a carrot or some celery, even a few slices of lemon or orange — things that will impart subtle and delicious flavors to your chicken. Use what you have on hand. If I'm grilling, I'll use molasses in place of sugar for its smoky, thick, taste. Honey and maple syrup have also made appearances. Experiment!

Tip: If you are in a hurry, cut back on the water when you are boiling the brine, then add cold water or ice to dilute the saltiness and cool it at the same time.

Poule au Pot

by Gordon Hamersley

What to do with an entire old bird? Something slow and delicious. This serves 6–8.

- 3 tablespoons olive oil
- 3 medium carrots, peeled and chopped
- 2 medium leeks, washed and chopped into 2-inch lengths
- 3 medium potatoes, peeled and cut into large pieces
- 6 cloves peeled garlic
- 2 sprigs thyme
- 2 bay leaves
- 1 sprig rosemary
- 1 tablespoon black peppercorns
- 2 dried chiles
- 1 tablespoon coriander seeds
- 1 tablespoon fennel seeds
- 1 bottle dry sherry

1 quart chicken stock
Salt
1 large old bird, about 6 pounds
Sprigs of flat-leaf parsley
Dijon mustard, for serving
Coarse mustard, for serving
Cornichons, for serving
Homemade pickles, for serving
Sea salt, for serving

In a large soup pot, heat olive oil. Add the vegetables and herbs and cook over medium heat for 6–8 minutes.

Add the sherry, chicken stock, salt to taste, and the chicken. Add enough water to just barely cover the bird.

Bring to a boil and then lower the heat to a simmer. Skim the foam as it rises to the surface. Cook over low heat until the chicken is tender, 40–50 minutes.

Lift the chicken out of the pot and let it rest for 5 minutes. Break the meat apart carefully, leaving the pieces large. Place on a large serving platter.

Using a large slotted spoon, arrange the vegetables around the chicken. Add the sprigs of parsley.

Pour the cooking liquor into a large water pitcher.

Take the platter of chicken, the pitcher of broth, and the condiments to the table. Give each person a shallow soup bowl, and let them serve themselves some chicken and vegetables. The meat is moistened with some of the broth. Condiments are added as each person wishes.

Taking Stock

Different stocks should be used for different kinds of cooking. The white stock, lighter in flavor, makes a great base for risottos, braised red meats or lamb, vegetable soups, or anything that you don't want to take on a strong chicken flavor. A roasted stock is more appropriate for things like chicken soup or stew, or even braised turkey legs, where you want a stronger chicken flavor and a darker color. If you don't have enough bones or scraps for these recipes, you can always freeze your ingredients until you have a good supply.

White Stock

5–10 pounds of raw chicken bones and pieces (necks, backs, wing tips, or bones removed from deboning a chicken)
Cold water

Place the chicken and bones in a large stockpot. Add cold water until it covers the chicken by about 3 inches. Put the pot on medium heat. As the water comes up to a low boil, remove any scum that forms on the top. Once the water starts to boil, turn down the heat to a simmer. Continue to cook at this temperature for 4 to 5 hours, skimming any scum from the surface.

Strain the stock into a large container and allow it to rest so the fat floats to the top. Skim off this fat, using a ladle, and save it (see Using Chicken Fat, below). The stock can be used right away, refrigerated for a week, or frozen.

Using Chicken Fat

Take the chicken fat conserved from making white stock and place in a saucepan, leaving at least 2 inches of room from the lip of the pot. Place over low heat and cook until any water has steamed off and the liquid stops bubbling. Let the fat cool slightly, then strain into a heatproof container and refrigerate.

RENDERING

To render the chicken fat, save chicken fat from whole birds and thighs or ask your butcher to save extra fat when breaking down chickens. Soak the chicken fat for an hour in cold water, then put it in a pan and cook over medium-low heat until most of the fat has rendered out and you can remove the chicken bits. Keep cooking until all the water has evaporated, and you'll be left with a wonderful cooking fat that resembles clarified butter. Store covered in the refrigerator. Be sure to save the fat that rises to the top when making white chicken stock.

Roasted Chicken Stock

5–7 pounds of chicken bones and pieces (necks, backs, wingtips, bones removed from a chicken or left over from a meal)
1 large carrot, peeled and trimmed
1 large onion, peeled and cut in half
1 stalk celery
Cold water

Preheat the oven to 400°F. Arrange all of the ingredients except the water in a single, not too tightly packed, layer on a baking tray or trays. Place in the oven and roast until the chicken pieces are all a deep brown, about 30 minutes. If any look close to burning, remove and set them aside until the others have finished. Pour off any fat that has been rendered by the roasting and discard.

Put all the roasted chicken and vegetables into a large stockpot. Use a wooden spoon and a bit of cold water to scrape all the browned bits from the bottom of the baking trays and into the pot. Cover the chicken and vegetables with cold water to about 2 inches above the bones. Put the pot on at a medium heat. As the stock comes to a boil, remove any scum that forms on the surface.

Once the pot is about to boil, turn the heat to low and simmer for 4 to 5 hours, continuing to scrape any scum from the surface. Strain the stock into a large container and use a ladle to remove any fat that collects at the surface. Discard this fat. This stock will keep for up to a week in the fridge or can be frozen.

Breaking Down and Deboning a Chicken

Breaking down chicken, whether splitting into multiple pieces or butterflying for the grill, is a fairly straightforward process. With a minimal amount of practice, you will feel like a pro.

Breaking a Chicken into Its Parts

Place the chicken breast-side up on a cutting board. Grab the end of the drumstick and pull it away from the body so that the skin between the thigh and the breast stretches. With a sharp knife, cut through this skin, favoring the thigh side. This will help to expose the joint where the thigh joins the body. Use your knife to free the leg from the body in the joint. At first this may seem complicated, but after a try or two it will be second nature.

Once the leg is free, if you wish to separate the thigh from the drumstick, lay the leg skin-side down on the cutting board and cut at the joint straight through the bend. Don't worry if you don't hit the joint perfectly; a chef's knife will easily cut through the top of the bone here. To debone the thigh, if you so desire, place the thigh face down on a cutting board. Imagine a line between the two exposed tips of bone. Drag the tip of your knife along this line. You should feel the knife following the bone. With a little trimming around the ends of the bone, the meat should come free. It's not really worth trying to debone a drumstick, as it has too many tendons. What you can do is cut in a straight line around the thin end and pull the meat down toward the thick end, creating a kind of lollipop. These are great grilled.

Once you have removed the legs, it's time to remove the wings. To do this, flip the bird over so that it is breast-side down. Grab the wing and wiggle it a bit to get a sense of where the wing joins the back. Make a semicircular cut to expose this joint, then cut through the joint. Flip the bird back over and finish cutting through the flesh. Leave as much meat on the body side of the cut as possible. You can separate the parts of the wing by simply pressing a chef's knife through the joints. Add the tips to your stock.

Now you have a bird with no wings and no legs. Removing the breast is fairly straightforward. For boneless breasts, start at the top of the bird and use the tip of your knife to cut down, riding first against the breastbone then the ribs, until the breast comes free. You will have to deal with the wishbone, which comes out into the breast slightly at the front of the chicken. You can either carefully cut around this or use your knife or shears to cut through it and remove it once you have the rest of the breast off.

For a bone-in breast, start by taking a sharp knife or kitchen shears and cut through the ribs, and wishbone from the back (large opening) to the front following the contour of the bottom of the breast meat. Save the back for stock. Once you have freed the breast from the back, place it breast-side down on the board and use a large kitchen knife to cut right through the middle of the breastbone. Or you can use a knife or kitchen shears to cut to either side of it. There is no one way to break down a chicken, but with a little practice it is easy to accomplish.

Butterflying Chicken

This is the best way to prepare a chicken for the grill. Place the chicken breast-side down on a cutting board. Using a sharp paring or boning knife, or, better yet, a pair of kitchen shears, cut away the body from either side of the spine and remove. I find if you are using a knife it works best to place the large opening away from you and work toward yourself (be careful), and the opposite if you are using shears.

Once you have the backbone removed (save it for stock), you have a choice. You can keep the chicken breast-side down and splay it open slightly. From here you can use your knife to cut through the soft cartilage on either side of the breastbone that runs down the middle of the chicken, and carefully remove the bone, or you can simply flip the chicken over, splay it out slightly, then press down. The chicken will flatten, breaking the bones in a clean line to one side of the breastbone.

Removing the breastbone is a little more "refined," but the choice is yours. Unless I'm cooking for someone I want to impress, I usually don't bother.

Grilling Chicken

1 chicken, 3–5 pounds, brined overnight and butterflied (see the brine recipe, page 112)
Salt and black pepper
About 2 cups of your favorite BBQ sauce (or make your own ahead of time; see next page)

Prepare a grill with medium, indirect heat. Charcoal is best, but gas will also work. Make sure your grill is well cleaned, and use a paper towel with a bit of olive oil on it and a pair of tongs to oil your grill surface. Sprinkle both sides of the chicken liberally with salt and pepper.

Place the chicken on the grill, breast-side up, cover, and cook for 45 minutes to an hour, depending on the size the chicken. Use a grilling brush to baste the bird with your sauce about every 15 minutes. Flip the chicken breast-side down directly over your heat source and baste the bone side of the bird liberally with more sauce. Cover and cook for another 5 to 15 minutes or until the skin is nice and crispy — slightly burnt.

BBQ SAUCE

- 1¾ cups cider vinegar
- ¼ cup tomato ketchup
- 1 tablespoon cayenne pepper
- 2 tablespoons molasses
- 1 garlic clove, minced
- 1 tablespoon black pepper
- 1 teaspoon salt, or to taste
- 1 slice of lemon

Combine all ingredients in a jar and mix well. Ideally this should sit for at least a week in the refrigerator before use. Any leftovers can be brought to a quick boil and used for dipping.

Poaching Chicken

by Robert Booz

- 1 tablespoon salt
- 1 heaping tablespoon of pickling spice tied in cheesecloth or a coffee filter
- 6–7 cups water
- 3 pounds boneless, skinless, chicken breasts, or a combination of breasts and boneless, skinless thighs
- 1 lemon, cut in half

Put the salt, pickling spice, and water in a large pot and bring it to a boil. Turn the heat to low and add the chicken, poaching for 20 minutes or until the chicken is firm and cooked through. Remove from the heat and add the two lemon halves, squeezing the juice out as you add them. Let all rest in the pot at least 30 minutes before removing the chicken. Discard all but the chicken.

Chicken Stew

by Daniele Dominick

- 5 pounds chicken legs, thighs, and breasts, skin removed (see Tip, page 119)
- Olive oil
- 1 large carrot peeled
- 1 stalk celery, whole
- 1 large Spanish onion, unpeeled
- 2 medium Spanish onions, chopped

4 stalks celery, cut into 1-inch pieces
4 large carrots, peeled, split lengthwise, and cut into 1-inch pieces
3 medium potatoes, cut into 1-inch pieces
1 bay leaf
1 6-ounce can of tomato paste
Salt and pepper

Season the chicken liberally with salt. In the bottom of a large stockpot, brown the chicken pieces with a bit of oil. Add the carrot, the whole celery stalk, and whole onion. Add water until all is covered by about 2 inches. Bring to a boil, then reduce to a simmer. Let simmer for about 3 hours.

Drain the vegetables and chicken, conserving the liquid. Slowly, over medium heat, reduce the stock by half (about 2 hours).

Shred the chicken and remove all the bones. In another pot, sauté the rest of the onions, celery, and carrot, in oil until nicely browned (about 10 minutes). Add the tomato paste and cook until it begins to darken and stick to the bottom of pot, being careful not to burn it. Add the shredded chicken, potatoes, and stock with bay leaf and salt and pepper to taste. Use a wooden spoon to scrape any remaining bits from the bottom of the pot. Add water to thin as needed.

Let the stew come to a boil, and then simmer for 30 minutes or until the vegetables and potatoes are cooked through.

TIP: The skin can be seasoned with salt, pepper, and smoked paprika and baked until crispy in a 350°F oven. It makes a wonderful substitute for bacon on a BLT.

Big Spiced Wings

by Gordon Hamersley

6 pounds large chicken wings (16 wings total)
5 tablespoons soy sauce
3 tablespoons cider vinegar
3 tablespoons roasted sesame oil
1 tablespoon salt
1 tablespoon five spice powder
2 teaspoons vanilla powder (a vanilla extract alternative)
1 tablespoon tamarind paste
1 tablespoon paprika
¼ cup honey
2 tablespoons cracked black pepper

(continued next page)

(Big Spiced Wings continued)
Combine all the ingredients but the chicken wings. Mix with the wings and marinate in the fridge for 30–60 minutes.

Lay the wings out on a rack fitted on a sheet pan. Brushing with marinade every 10 minutes, roast at 375°F for 35–40 minutes. Serve hot.

It's Offal

The innards are the best parts. The next five recipes are by Jefferson Munroe, and the last is by Betsy Carnie.

Korean BBQ Hearts

Grilled to chewy perfection, this serves 6 as an appetizer.

- 1 pound chicken hearts
- 1 1-inch knob of ginger root, grated
- 1 garlic clove, finely chopped
- ¼ cup chili paste (sriracha or sambal)
- ¼ cup molasses
- 1 tablespoon soy sauce
- 1 teaspoon black pepper

Clean the chicken hearts of any residual blood or membranes. Combine all ingredients but the hearts in a nonreactive bowl, mixing well. Add the hearts, being sure to coat them thoroughly. Cover and refrigerate for at least 3 hours or preferably overnight.

Skewer the hearts and grill on the top rack in a covered grill over medium heat for 7 minutes on each side or until the outside begins to caramelize and the hearts firm up. If you are using a charcoal grill, be sure that the coals are at least 4 inches below the grill surface.

Moroccan Braised Gizzards

Inspired by a beloved Paula Wolfert recipe, this tagine can cook and cook and cook!

2 pounds chicken gizzards, well cleaned
2 tablespoons extra-virgin olive oil or chicken fat
1 large white onion, chopped
2 garlic cloves, chopped finely
1 fennel bulb, chopped finely
2 stalks rhubarb, chopped finely
2 teaspoons ground cumin
½ teaspoon ground ginger
½ teaspoon ground cinnamon
½ teaspoon cayenne pepper
½ teaspoon turmeric or a pinch of saffron
2 cups water or stock
1 15-ounce can chickpeas
1 tablespoon apple cider vinegar
½ cup kalamata or green olives, pitted
3 tablespoons chopped cilantro, for serving
Salt and pepper, to taste
Hot rice, for serving

Preheat the oven to 220°F. Clean the gizzards and wash thoroughly. Sprinkle with salt. Heat oil over medium-high heat and brown the gizzards. Be sure to use a heavy-bottomed pan and brown only a few at a time to ensure proper browning. Once they're browned, remove them to a separate dish and lower the heat to medium. Sauté the onion, garlic, and fennel until softened and lightly browned, about 10 minutes. Add the rhubarb, cumin, ginger, cinnamon, cayenne, and turmeric, and and sauté for another 5 minutes; then add the water, chickpeas, vinegar, and the gizzards.

Once the mixture comes to a simmer, cover and place in the oven. Cook for 2 hours. Give it a stir after the first hour. After 2 hours, check to make sure the gizzards are tender, adjust the seasoning, and add the olives. Let the whole pot rest for 10 minutes, then serve over rice with the chopped cilantro.

Confit of Gizzards

Confit, a term that comes from French, refers to something that has been cooked in fat, but it might as well mean "deliciousness cooked in delicious fat" because it makes everything better. These gizzards should be put away for a week in your fridge after cooking to let the flavors meld. Then they're great browned with some eggs in the morning (think unsmoked bacon) or cooked under the broiler for 7 minutes before tossing with some greens, fresh herbs, and vinegar. Or without anything else whenever you're in the mood for a tasty little snack. Properly stored they'll keep up to three months in your fridge, but I doubt that they'll last that long.

- 2 pounds gizzards
- 4 garlic cloves
- 2 bay leaves
- 3 sprigs thyme
- 3 tablespoons kosher salt
- 1–2 cups rendered chicken fat, duck fat, or lard (in a pinch you can use olive oil)

Clean and thoroughly wash the gizzards. Dry well; curing is intended to remove as much water as possible. Smash the garlic, break up the bay leaves, pluck the leaves from the thyme sprigs, and toss with the gizzards and salt in a bowl. Place in fridge and leave overnight. Drain off any excess liquid in the morning and place the gizzards in a colander over a bowl. Wait another 6 hours.

Preheat the oven to 220°F. Pat dry the gizzards, removing any seasoning that has clung to them. Combine the gizzards with the fat in an ovenproof pan and place over medium-low heat. Once the fat reaches a simmer, move it to the oven and allow to cook for 4 to 6 hours or longer. You'll know the gizzards are done when they're lying at the bottom of the pan and the fat looks clear, as this will indicate that they are no longer releasing any liquid. To store I usually use Mason jars, but any container will do. Fill the container four-fifths full with gizzards, then pour the melted fat over them, making sure not to scoop any of the cooking liquid from the bottom of the pan as it will spoil too quickly. The fat should completely cover the gizzards. Be sure there are no air pockets. Any extra cooking liquid works well as a concentrated, salty stock for cooking grains or adding to vegetable sautés.

Chicken Foot Souse

Best eaten with your hands, this makes perfect (and exciting!) picnic fare.

- 1 lb chicken feet
- 3 lemons
- 1 head of garlic, unpeeled
- ¼ cup salt plus more to taste
- 1 small bunch cilantro
- 2 pimiento peppers
- 1 medium red onion, peeled, halved, sliced, and separated into half-rings
- 1 cucumber, peeled and sliced into rounds
- 1 hot pepper such as habañero or jalepeño, minced (optional)

Wash the feet, and with a sharp paring knife trim away the toes and any dirt that has been impacted into the soles. Cut one lemon in half and rub the halves all over the feet, paying special attention to any openings in the skin. Place the feet in a pan with the garlic and cover with water. Add the ¼ cup of salt and bring the pot to a simmer; simmer for 2 hours or until tender. Meanwhile, juice the remaining lemons and set aside the juice. Once the feet are tender, drain and rinse them three times in fresh, cold water.

Chop the cilantro and pimiento peppers and puree in a food processor, adding salt to taste. Place the feet, onion, and cucumbers in a bowl and pour 1 cup water, the lemon juice, the cilantro mixture, and the hot pepper, if using, over the feet. Mix well and allow to marinate for 3 hours.

Chicken Liver Mousse

- 1 pound chicken livers, cleaned of any connective tissue and soaked overnight in milk
- 1 tablespoon chicken fat or extra-virgin olive oil
- 1 stick butter cut into 5 pieces, at room temperature, plus 1 tablespoon
- 1 medium Spanish onion, peeled, halved, sliced and separated into half-rings
- 2 cloves garlic, thinly sliced
- Salt and freshly ground black pepper
- ¼ cup Madeira
- 1 pinch ground pistachios
- 1 sprinkle unsweetened cocoa powder

(continued next page)

(Chicken Liver Mousse, continued)

Rinse the livers and dry well. Place a pan over medium-high heat and add the chicken fat or olive oil. Once the pan is good and hot, sear the livers in small batches, browning well. Set the livers aside, reduce the heat to medium and add the 1 tablespoon of butter and the onions and garlic saled to tast. Cook until softened and lightly browned. Deglaze the pan with the Madeira and add the livers back. Cook, stirring, until the livers are cooked through, about 3 minutes.

Remove the livers from the heat and let them cool to room temperature, which will take about 20 minutes. You want the butter at room temperature and the livers and onion in the same range — if the livers are too hot, they'll melt the butter and your mousse will separate; if the butter is too cold, the food processor will cut it up rather than whip it into a mixture. It'll still be delicious, but the texture won't be quite the same.

When everything is at the same temperature, put the liver mixture in the food processor and get it whizzing away until it is well puréed, then add the butter to the spinning processor one piece at a time. When the butter has been incorporated, season again with salt and pepper, keeping in mind that when cold you won't taste the salt as much. Scoop it into a suitable container and sprinkle with ground pistachios and a dusting of cocoa powder. Eat with just about anything crunchy.

Chicken Liver Pâté

created by Betsy Carnie

- 5 large chicken livers, about 7 ounces, soaked in milk overnight
- 2 tablespoons chicken fat or clarified butter
- ½ small yellow onion, thinly sliced
- ½ cup sherry
- 1 teaspoon herbes de Provence
- 1 teaspoon fresh lemon thyme, minced, plus a sprig for serving
- ½ teaspoon freshly ground nutmeg
- ¼ teaspoon ground cloves
- 1 teaspoon salt or to taste
- 1 teaspoon freshly ground black pepper
- 1 8-ounce package Neufchatel cheese, at room temperature

Rinse and dry the chicken livers.

Add the chicken fat or clarified butter to a pan over medium-high heat. When the pan is hot, brown the chicken livers well; reduce the heat to medium. Add the onion and sauté until lightly browned, approximately 5 minutes. Deglaze the pan with the sherry, scraping any bits from the bottom. Transfer the contents of the pan into the bowl of a food processor and add the herbs, spices, salt, and pepper. Puree until smooth, then add the Neufchatel cheese in 2-ounce pieces while the food processor is running. Puree until well incorporated. Adjust seasoning to taste. Use a spatula to turn and the press pâté into a tureen or ramekin of your choice. Or you can line a mold with plastic wrap and press the pâté into that. Chill overnight.

Unmold it to serve. Garnish with a fresh thyme sprig and serve with thinly sliced baguette or crostini and a couple of crackers.

Comfort Food

Eating at the Scottish Bakehouse on Martha's Vineyard means eating a square meal in every way: flavor, clean fair food, and balance. Owner and cook Daniele Dominick manages to provide affordable meals (mostly to go) using local produce, local chicken, and as many organic ingredients as she can manage. If she makes less on the local food, she balances it out with the other less-expensive ingredients.

Local chicken looks, tastes, and cooks differently, Daniele says.

"I just jumped in with the local chicken once it was available," she says. "I had to. If you just look at the cost of local meat on paper, you'll never do it, as a business owner. But once you get your hands on it, you'll see the difference."

The way she makes local chicken affordable is by using the whole bird and knowing her clientele. They are working people: builders, landscapers, window washers. Feeding them hearty, home-cooked, and comforting foods that will fill them up without putting them to sleep is Daniele's mission. Her chicken stew (see page 118), for example, is an economical way to use the whole bird, and it hits her customers' sweet spot. Comfort food.

Daniele's advice to a restaurant that wants to start using the more expensive local chicken: "Know your customers. Buy only a small quantity of birds and go from there. Start with a Special.

"Local chicken looks, tastes, and cooks differently," she adds. "If all you know is Sysco and you never try local chicken, you don't know what the differences are."

Appendix

FARMER SURVEY SUGGESTIONS

Find out what's going on in your farming community by taking a survey — perhaps during an off-season farmer's dinner. Here are some sample questions.

- ☐ Do you raise poultry now? How many egg layers? How many broilers? What breeds?
- ☐ If you don't raise broilers, why not?
- ☐ If you do raise broilers for sale, where do you take them to be slaughtered?
- ☐ Are you satisfied with the outcome? Consider:
  - confidence in animal handling
  - disposal of waste
  - quality of product (appearance, taste, texture)
  - cost of slaughter, processing
- ☐ How much do you think an eater will pay for local chicken?
- ☐ How much do you charge per pound for your chicken?
- ☐ Do you pasture your birds?
- ☐ What do you describe as your "chicken season"? (For example, is it April to November?)
- ☐ When do you order your chicks? From which hatchery do you order your chicks?
- ☐ How many do you raise at a time? How many would you raise if you could?
- ☐ Do you raise other animals for meat?
- ☐ If a there was a licensed mobile trailer or unit, would you use it on your farm?
- ☐ If a mobile trailer or unit was available to you, would you raise more broilers?
- ☐ Do you compost? Yes / No
- ☐ Are there any speakers you would like to hear or topics specifically to address?

FARMER'S CHECKLIST FOR DAY OF

Adapt this checklist to your farm, the farmers you serve, and your community. Determine ahead of time how water-soaked woodchips, all INEDIBLE offal, and used chill-tank water will be transported to compost.

- ☐ Number of birds slotted for humane slaughter: _____
- ☐ Date to be slaughtered: ________________
- ☐ Any special considerations (weather, age of birds, other…): _____________

  ____
- ☐ What number in order of slaughter (first, second, third, etc.)?: ______
- ☐ Chicken Crew contact info:

  1)

  2)

  3)
- ☐ Site: The site must be above and away from the water table. Do not set up near a pond or stream, and look for some (not much) elevation so the wastewater drains away from the working area. This site should be a level workspace, grass or cement, about 15' x 25', free of all fire hazards and spacious enough for two tents.
- ☐ Recommended: Hot water source nearby to fill scalder and reduce the time it takes for scalder to heat
- ☐ Electricity: 20-amp breaker on a 100-amp 120-volt service
- ☐ Two working, safe extension cords
- ☐ Two food-grade hoses with spray attachments for potable water
- ☐ One 4-way hose splitter
- ☐ Copy of the farm's *Potable Water* certification for the Crew manager and/or regulator
- ☐ ½–1 cubic yard of clean wood chips (not pine shavings or sawdust)
- ☐ Safe holding crates for birds
- ☐ Caged healthy birds, feed withheld for at least 8 to 10 hours, preferably 12
- ☐ Clean, loose, cubed ice: 4–5 pounds/bird (err on the side of more ice)
- ☐ 4 plastic chill tanks with lids, 30 gallons minimum, marked with indelible ink: *EDIBLE*
- ☐ A drain rack for dressed birds
- ☐ Labels for bagged birds
- ☐ Refrigeration space ready for processed birds: 34°F with a thermometer in place
- ☐ Compost that includes wood chips, ready and accessible for inedibles
- ☐ Fire extinguisher on hand and functional
- ☐ First-aid kit accessible and up to date
- ☐ Meet the Chicken Crew when they arrive and check in throughout the processing.
- ☐ Have payment ready for the Crew.

Resources

ESSENTIAL ORGANIZATIONS

Farm-Based Education Association
www.farmbasededucation.org

Farm-to-Consumer Legal Defense Fund
www.farmtoconsumer.org
Protects the rights of family farms, artisan food producers, consumers, and affiliate communities to engage in direct commerce free of harassment by federal, state, and local government interference

Farmers Market Coalition
http://farmersmarketcoalition.org

National Restaurant Association Educational Foundation
www.servsafe.com
ServSafe Food Safety certification classes

Niche Meat Processor Assistance Network
www.nichemeatprocessing.org
For updates on slaughter programs, discussions about the nitty-gritty, and informative webinars regarding mobile slaughter units for four-legged and poultry.

Slow Food USA
http://slowfoodusa.org
A global, grassroots movement with thousands of members in over 150 countries, which links the pleasure of food with a commitment to community and the environment.

Transition Network
www.transitionnetwork.org
Supports community initiatives that rebuild resilience and reduce CO_2 emissions.

FUNDING

There is a trend in Public Health and the Centers for Disease Control to connect to local food systems with grant monies. Check in with your state's public health departments for funding opportunities.

Find the USDA Rural Development agencies and offices near you. Google Hint: USDA Rural Development regional offices. Ask your regional office to come and speak to your group.

Foundation Center
www.foundationcenter.org
National foundation research

Sustainable Agriculture & Food Systems Funders
www.safsf.org
If you're a philanthropist or grantmaker interested in supporting sustainable agriculture, use this site to find and follow trends

Sustainable Agriculture Research & Education (SARE)
www.sare.org
Download this SARE book for free: Building Sustainable Farms, Ranches and Communities: Federal Programs for Sustainable Agriculture, Forestry, Entrepreneurship, Conservation, and Community Development

PRACTICAL SKILLS AND EQUIPMENT

UMass Extension, Center for Agriculture
www.extension.org/pages/24718/alternatives-to-rendering:-butcher-waste-composting
Alternatives to Rendering: Butcher Waste Composting Information

Cornell Waste Management Institute
http://cwmi.css.cornell.edu/factsheets.htm
Composting fact sheets

Cornerstone Farm Ventures
www.cornerstone-farm.com
Processing equipment, supplies, information, workshops, and consulting. The sole distributor of Poultryman products. Poultryman equipment is made by Old World Mennonites; hence, not on the Internet, but shipped all over the world.

Homestead Poultry Butchering
www.themodernhomestead.us/article/Butchering-Evisceration-1.html
How to eviscerate a chicken

Scale Labels Etc.
www.scalelabelsetc.com/shi
Safe Handling stickers

BOOK LIST

Charles, Prince of Wales. *The Prince's Speech: On the Future of Food.* Rodale, 2012.
Damerow, Gail. *Storey's Guide to Raising Chickens,* 3rd ed. Storey Publishing, 2010.
Denckla Cobb, Tanya. *Reclaiming our Food.* Storey Publishing, 2011.
Eastman Jr., Wilbur F. *A Guide to Canning, Freezing, Curing & Smoking Meat, Fish & Game,* rev. ed. Storey Publishing, 2002.
Ekarius, Carol. *Storey's Illustrated Guide to Poultry Breeds.* Storey Publishing, 2007.
Gawande, Atul. *The Checklist Manifesto.* Metropolitan Books, 2009.
Grandin, Temple. *Animals Make Us Human.* Houghton Mifflin Harcourt, 2009.
_____. *Animals in Translation.* Houghton Mifflin Harcourt, 2006.
_____. *Thinking in Pictures.* 2nd ed. Vintage Books, 2006.
Horowitz, Roger. *Putting Meat on the American Table.* John Hopkins University Press, 2006.
Kessler, David A. *The End of Overeating.* Rodale, 2009.
Kirschenmann, Frederick L. *Cultivating an Ecological Conscience.* University Press of Kentucky, 2010.
Safran Foer, Jonathan. *Eating Animals.* Little, Brown & Co., 2009.
Schumacher, E. F. *Small Is Beautiful.* Harper Perennial, 2010.
Sinclair, Upton. *The Jungle.* Dover, 2001. First published 1906.
Wendell, Berry. *The Unsettling of America,* rev. ed. Sierra Club Books, 1997.

For the activist and the advocate:

Dernoot Lipsky, Laura van, and Connie Burk. *Trauma Stewardship.* Berrett-Koehler Publishers, 2009.
Hauter, Wenonah. *Foodopoly: The Battle Over the Future of Food and Farming in America,* The New Press, 2012.
Nestle, Marion. *Food Politics.* University of California, 2002.
_____. *What to Eat.* North Point Press, 2006.
Planck, Nina. *Real Food.* Bloomsbury USA, 2006.
Pollan, Michael. *Food Rules.* Penguin, 2009.
_____. *In Defense of Food.* Penguin, 2008.
_____. *The Omnivore's Dilemma.* Penguin, 2006.
Roberts, Paul. *The End of Food.* Houghton Mifflin, 2008.
Salatin, Joel. *Holy Cows & Hog Heaven.* Polyface, 2005.
Schlosser, Eric. *Fast Food Nation.* Houghton Mifflin, 2001.
Scully, Matthew. *Dominion.* St. Martin's Griffin, 2002.
Serpell, James. *In the Company of Animals.* Cambridge University Press, 1996.
Singer, Peter, and Jim Mason. *The Way We Eat.* Rodale, 2006.
Squier, Susan M. *Poultry Science, Chicken Culture.* Rutgers University Press, 2011.
Striffler, Steve. *Chicken: The Dangerous Transformation of America's Favorite Food.* Yale University, 2005.

Cookbooks:

Ash, John. *John Ash: Cooking One on One.* Clarkson Potter, 2004.
Child, Julia, Louisette Bertholle, and Simone Beck. *Mastering the Art of French Cooking,* 40th anniv. ed. 2 vols. Knopf, 2011.
Hamersley, Gordon. *Bistro Cooking at Home.* Broadway Books, 2003.
Jones, Judith. *The Pleasures of Cooking for One.* Knopf, 2009.
Oliver, Jamie. *Jamie's Food Revolution.* Hyperion, 2009.
Rombauer, Irma S., Marion Rombauer Becker, and Ethan Becker. *Joy of Cooking,* 75th anniv. ed. Scribner, 2006.
Ruhlman, Michael, and Brian Polcyn. *Charcuterie: The Craft of Salting, Smoking, and Curing.* W. W. Norton, 2005.
Stevens, Molly. *All About Roasting; A New Approach to a Classic Art.* W. W. Norton & Company, 2011.
Waltuck, David, and Melicia Phillips. *Staffmeals from Chanterelle.* Workman Publishing, 2001.
Wolfert, Paula. *Mediterranean Clay Pot Cooking.* John Wiley & Sons, 2009.

In the spirit of "eat less or no meat":

Colbin, Annemarie. *Food and Healing,* 10th anniv. ed. Ballantine Books, 1986.
Lappé, Frances Moore. *Diet for a Small Planet,* 20th anniv. ed. Ballantine Books, 1991.
Pitchford, Paul. *Healing with Whole Foods.* North Atlantic Books, 1993.
Swanson, Heidi. *Super Natural Every Day.* Ten Speed Press, 2011.

Index

Note: All recipes are indexed under "recipes."

Other Storey Titles You Will Enjoy

The Chicken Health Handbook, by Gail Damerow.
A must-have reference to help the small flock owner identify, treat, and prevent diseases common to chickens of all ages and sizes.
352 pages. Paper. ISBN 978-0-88266-611-2.

Greenhorns, edited by Zoë Ida Bradbury, Severine von Tscharner Fleming, and Paula Manalo.
Fifty original essays written by a new generation of farmers.
256 pages. Paper. ISBN 978-1-60342-772-2.

The Organic Farming Manual, by Ann Larkin Hansen.
A comprehensive guide to starting and running, or transitioning to, a certified organic farm.
448 pages. Paper. ISBN 978-1-60342-479-0.

Reclaiming Our Food, by Tanya Denckla Cobb.
Stories of more than 50 groups across America that are finding innovative ways to provide local food to their communities.
320 pages. Paper. ISBN 978-1-60342-799-9.

Storey's Guide to Raising Chickens, by Gail Damerow.
The ultimate guide that includes information on training, hobby farming, fowl first aid, and more.
448 pages. Paper. ISBN 978-1-60342-469-1.
Hardcover. ISBN 978-1-60342-470-7.

Storey's Guide to Raising Poultry, by Glenn Drowns.
An invaluable resource of essential information on housing, breeding, and caring for chickens, turkeys, ducks, geese, guineas, and game birds.
464 pages. Paper. ISBN 978-1-61212-000-3.
Hardcover. ISBN 978-1-61212-001-0.

These and other books from Storey Publishing are available wherever quality books are sold or by calling 1-800-441-5700.
Visit us at *www.storey.com* or sign up for our newsletter at *www.storey.com/signup.*